Sensors 2019

Topical Conference at the 2019 AIChE Annual Meeting

Orlando, Florida, USA
10-15 November 2019

ISBN: 978-1-7138-0558-8

Printed from e-media with permission by:

Curran Associates, Inc.
57 Morehouse Lane
Red Hook, NY 12571

Some format issues inherent in the e-media version may also appear in this print version.

AIChE
120 Wall Street, FL 23
New York, NY 10005-4020

Phone: (800) 242-4363
Fax: (203) 775-5177

www.aiche.org

Additional copies of this publication are available from:

Curran Associates, Inc.
57 Morehouse Lane
Red Hook, NY 12571 USA
Phone: 845-758-0400
Fax: 845-758-2633
Email: curran@proceedings.com
Web: www.proceedings.com

Sensors 2019

Topical Conference at the 2019 AIChE Annual Meeting

Orlando, Florida, USA
10-15 November 2019

AIChE
The Global Home of Chemical Engineers

TABLE OF CONTENTS

576740 Multifunctional Hydrogel-Coated Gold Nanoshells for Protein Biomarker Quantification

There has been growing interest in using biomarkers expressed in readily available patient samples, such as saliva, tears, urine, or serum, to develop minimally-invasive tests for disease screening and diagnosis. While many different approaches have been explored, differential sensors exploiting arrays of semi-selective recognitive elements are a promising strategy for extending current methods to multiple marker quantification without the need for the development of a new receptor for each biomarker. However, translating differential sensing approaches to protein recognition is challenged by the similarity of the biomarkers' structures and properties to those of the other proteins present in the sample. In order address this challenge, we explore the capabilities of hydrogel network formation on the surface of gold nanoshells (AuNS) as a semi-selective recognition element for the quantification of protein biomarkers based on differences in polymer functionalization using shifts in the maximum localized surface plasmon resonance (LSPR) wavelength.

AuNS were prepared by seeded growth of colloidal gold on the surface of aminated silica nanoparticles followed by encapsulation with poly(maleic anhydride-*alt*-1-octadecene)-g-poly(ethylene glycol) methacrylate (PMAO-g-PEGMA) graft copolymer to improve particle stability. Hydrogel shells were formed by incorporating the AuNS into the aqueous phase of an inverse emulsion copolymerization of methacrylic acid with acrylamide or n-isopropylacrylamide, crosslinked with N,N'-methylenebis(acrylamide). Transmission electron microscopy was used to determine nanoshell size distribution and UV-Vis spectroscopy was used to measure the shifts in the wavelength of the LSPR. These polymer-coated AuNS can serve as a semi-selective sensor for protein adsorption, which will produce a red shift in the LSPR, and can be modified by incorporating comonomers with selective functionalities or by post-synthesis modification through carbodiimide coupling in order to improve the selectivity of the sensor for certain biomarkers.

Work supported by National Institutes of Health Grant R01-EB022025

574741 Surface Effects of Ultrasmall Nanoparticle on Cellular Uptake, Proliferation, and Multipotency of Neural Stem Cells

Ultrasmall iron oxide nanoparticles (USIONPs) have been recently developed as novel labelling probes as positive contrast agents for magnetic resonance imaging (MRI). However, their use in stem cell tracking have not been reported. Here, we studied the effects of USIONP surface coatings on proliferation, cellular uptake, and multipontency of established neural stem cells (NSCs) and primary NSCs. USIONPs were functionalized with gluconic acid (GA), tannic acid (TA), and hyaluronic acid (HA) to label NSCs. All functionalized USIONPs were characterized as T1 contrast agents via relaxivity measurements. Nanoparticles functionalized with TA and HA-coating promoted NSC proliferation and enhanced cellular uptake in a dose-dependent manner. Furthermore, HA-coating showed enhanced cell proliferation and cellular uptake in primary NSCs depending on the HA molecular weight. Stem cell characteristics were well-maintained, verified by neurosphere formation and gene expression levels of stemness and differentiation markers. Collectively, we demonstrated that NSC proliferation, cellular uptake, and multipotency can be enhanced using various surface coating strategy of USIONPs.

575596 Quantum Capacitance Based Amplified Graphene Phononics for Studying Neurodegenerative Diseases

Amyotrophic lateral sclerosis (ALS) is the most common adult-onset motor neuron disease, characterized by a rapid loss of upper and lower motor-neurons resulting in patient death from respiratory failure within 3-5 years of initial symptoms onset. Although at least 30 genes of major effect have been reported, the pathobiology of ALS is not well understood. Compounding this is the lack of a reliable laboratory test which can accurately diagnose this rapidly deteriorating disease. Herein, we report on graphene's phonon vibration-energies as a sensitive measure of the composite dipole moment of the components of the interfaced cerebrospinal fluid (CSF) to specifically identify patients with ALS disease. The second-order overtone of in-plane phonon vibration energy (2D) of graphene shifts by 3.2 ± 0.5 cm^{-1} for all ALS patients studied in this work. Further, the amount of *n*-doping induced shift in phonon energy of graphene, interfaced with CSF, is specific to the investigated neurodegenerative disease (ALS, Multiple Sclerosis and Motor Neuron Disease). By removing a severe roadblock in disease detection, this technology can be applied to study diagnostic biomarkers for researchers developing therapeutics and clinicians initiating treatments for neurodegenerative diseases.

567943 Rare Earth Element Doped Nanocrystals for Correlative Cathodoluminescence Electron Microscopy Bioimaging (CCLEM)

Correlative imaging has enabled some of the biggest scientific discoveries in the past. It has become a powerful analytical tool for unravelling structure-function relationships. However, correlating optical and electron microscopy is often challenging due to the resolution mismatch and different sample preparation requirements.[1] Interestingly, accelerated electrons can also generate luminescence signal, creating an effect called cathodoluminescence (CL)[2, 3]. Correlative cathodoluminescence electron microscopy (CCLEM) bioimaging correlates cathodoluminescence and electron microscopy images[4]. Although luminescence excitation by an electron beam enables sub-diffraction imaging, structural damage to the sample by high energy electrons has been identified as a potential obstacle.

Here, we present rare earth (RE) element doped nanocrystals as bright and stable probes for high-quality CCLEM. We report on the performance of RE-doped nanocrystals in CCLEM compared to fluorophores and quantum dots commonly used as molecular labels in biology. The influence of the electron beam parameters on image quality and resolution will be presented along with the first experimental demonstration of correlative cathodoluminescence back-scattering electron microscopy on focused ion beam (FIB)-sectioned biological samples[4]. We also demonstrate how exogenous nanoparticles can be easily distinguished from naturally occurring cellular features based on the characteristic optical emission. Taken together, we demonstrate the potential of the CCLEM approach for the acquisition of high-quality luminescence and electron microscopy images with nanometric resolution of biological samples.

1. Hell, S. W.; Sahl, S. J.; Bates, M.; Zhuang, X.; Heintzmann, R.; Booth, M. J.; Bewersdorf, J.; Shtengel, G.; Hess, H.; Tinnefeld, P.; Honigmann, A.; Jakobs, S.; Testa, I.; Cognet, L.; Lounis, B.; Ewers, H.; Davis, S. J.; Eggeling, C.; Klenerman, D.; Willig, K. I.; Vicidomini, G.; Castello, M.; Diaspro, A.; Cordes, T. *Journal of Physics D: Applied Physics* **2015,** 48, (44), 443001.
2. Mahfoud, Z.; Dijksman, A. T.; Javaux, C.; Bassoul, P.; Baudrion, A.-L.; Plain, J.; Dubertret, B.; Kociak, M. *The Journal of Physical Chemistry Letters* **2013,** 4, (23), 4090-4094.
3. Atre, A. C.; Brenny, B. J. M.; Coenen, T.; García-Etxarri, A.; Polman, A.; Dionne, J. A. *Nature Nanotechnology* **2015,** 10, 429.
4. Keevend, K.; Stiefel, M.; Neuer, A. L.; Matter, M. T.; Neels, A.; Bertazzo, S.; Herrmann, I. K. *Nanoscale* **2017,** 9, (13), 4383-4387.

568345 Effect of Particle Diameter and Magnetic Anisotropy on Magnetorelaxometry and Magnetic Particle Imaging Performance of Immobilized Magnetic Nanoparticles

Magnetic nanoparticles (MNPs) are of interest for application in magnetic particle imaging (MPI), an emerging biomedical imaging technology that relies on the non-linear dynamic magnetization response of MNPs. Compared to other molecular imaging modalities, MPI has the advantages of not requiring ionizing radiation and having high image contrast (zero signal from the tissue background), high sensitivity, and zero signal depth attenuation. In MPI, a static bias magnetic field gradient is superimposed with a uniform alternating magnetic field (AMF) to generate a small "field-free region", where the particles are able to respond to the AMF and generate a response signal. This signal is monitored by a pick-up coil and is then used to generate a quantitative image of the distribution of MNPs in a field of view. The variation of signal intensity with distance in the space around a point source in the imaging volume is commonly described using the so-called point spread function (PSF), which provides information on the signal strength and theoretical resolution obtainable with a given nanoparticle. A review of the recent literature suggests that no prior computational work has studied the effect of nature of magnetocrystalline anisotropy, strength of magnetocrystalline anisotropy energy, and particle diameter on the MPI performance (through the point spread function) of magnetic nanoparticles undergoing Néel-relaxation. On the other hand, prior computational studies have applied simulations of the Landau-Lifshitz-Gilbert (LLG) equation to predict the magnetorelaxometry of MNPs. However, these did not provide insight into the dipole dynamics for nanoparticles that are fixed in a physical matrix, such as would often be the case for nanoparticles that accumulate inside cells. Thus, further work is needed to fully understand how nanoparticle properties influence non-linear magnetization dynamics of nanoparticles, particularly in the context of the fields often used in MPI.

Here we report a computational study of the effect of particle diameter and magnetic anisotropy (considering both symmetry and energy) on the magnetization dynamics and MPI performance of spherical single-domain MNPs that relax by internal dipole rotation, by carrying out simulations that are based on the LLG equation. In the case of an applied static magnetic field, we studied the magnetic relaxation time of the nanoparticles as the field is suddenly applied and suppressed, for particles with uniaxial and cubic magnetic anisotropies and various anisotropy constants and particle diameters. The results, for both anisotropy symmetries show that for both cases where a static magnetic field is suddenly turned on or off, the MNPs may undergo a successive two-step (periods τ_1 and τ_2 for field turned on, periods τ_3 and τ_4 for field turned off) or combined one step (period τ_{12} for field turned on, period τ_{34} for field turned off) relaxation. Whether a particle relaxes with one or two periods when the field is turned on is independent of the particle size, but determined by the competition between the energy due to the applied magnetic field and the magnetic anisotropy energy. For the case where the applied magnetic field is suddenly turned off, our results show good agreement with theoretical predictions for the cases of $|K|V/k_B T \leq 1$ and $|K|V/k_B T >> 1$, where K represents the magnetocrystalline anisotropy constant, V represents the particle volume, k_B is the Boltzmann constant and T represents the absolute temperature. On the other hand, for the case of a dynamic applied AMF that is typical of MPI applications, the magnetic responses of the magnetite nanoparticles of various diameters were studied in terms of the magnetization dynamics, dynamic hysteresis loops, harmonic spectra, and PSF of the magnetization signal as a function of particle diameter. Results suggest that increasing the particle diameter leads to improved MPI performance (signal and resolution). Comparison of the MPI performance was also made between nanoparticles with uniaxial and cubic anisotropies. In summary, these computational studies provide theoretical insight into the role of particle diameter and magnetic anisotropy on the performance of MNPs for applications in MPI.

569135 DNA-Caged Polymer Nanocomposites for Erasable Fluorescence Imaging

Purpose: DNA follows well-known base pairing rules that allow for the creation of novel structures, including DNA origami. DNA origami can be designed into a wide variety of two- and three-dimensional shapes for a variety of applications, such as nanoscale engineering and drug delivery. Here, we report DNA cages, DNA nanostructures designed to form a network on surfaces. This DNA nanostructure was originally designed by Kurokawa et al.[1] for use as an artificial cytoskeleton inside liposomes. As a cytoskeleton, DNA cages increased the mechanical integrity of the lipids. Our goal is to integrate DNA cages with known micelle structures formed from block copolymers for applications in imaging, including erasable labeling of biological samples. Current procedures for multiplexed imaging via erasable labels require harsh treatment of tissue that can damage the samples. Multiplexed images allow for the labeling of multiple targets simultaneously with high image resolution, which means that complicated pathways in samples such as tumors can be studied using less time and materials. If the samples are damaged before all desired parts are labeled, additional samples may be required, which is not always possible when the samples come from a patient. Further, sequential sample labeling precludes evaluating the same markers in the same cells. The proposed labeling system is based on gentle DNA dehybridization interactions, which would greatly improve current technology.

Methods: The DNA origami previously described[1] was introduced to polymer micelles and solid polymer nanoparticles, forming a cage on their surfaces. Previously[1], it was suggested that DNA interacts with the surface of the micelles based on the charged group on the surface of the micelle and the negatively-charged DNA strands. Here, we show that attraction may also be based on the affinity of DNA for the polymer used. Proof of cage formation was measured based on the fluorescence of the cage itself, until a saturation point on the surface of the micelles was reached. DNA cages were tagged with single stranded DNA (ssDNA) targeting strands that reversibly attach to ssDNA strands labeled with fluorophores via strand displacement reactions. Strand displacement reactions are based on the highest number of complimentary base pairs. Thus, when an ssDNA with greater complement was added, the labeling strand was removed, erasing the signal.

Results and Implications: These data suggest that DNA cages have great promise for multiplexed cell labeling. Proper diagnosis of patients, especially when studying tumors, is highly dependent on cell labeling. DNA cages do not require harsh chemicals or procedures for removal, which may improve tissue labeling. The erasable nature of DNA cages allows for the labeling of the same target multiple times or different targets sequentially, potentially without damaging the sample. Finally, when combined with advanced image analysis techniques, it may be possible to create a three-dimensional image of the sample for a more complete analysis. Thus, DNA cages have the potential to significantly change cell labeling, especially that of patient samples.

Reference

1 Kurokawa, C. *et al.* DNA cytoskeleton for stabilizing artificial cells. *Proc Natl Acad Sci U S A* **114**, 7228-7233, doi:10.1073/pnas.1702208114 (2017).

572824 Tuning the Size and Composition of Manganese Oxide Nanoparticles for Magnetic Resonance Imaging of Breast Cancer

Magnetic resonance imaging (MRI) can detect more breast cancers than mammography, the current gold standard screening technique. However, MRI still produces a high false positive rate of up to 25% due to the commonly used gadolinium contrast agent, which is always active and nonspecifically highlights all vascularized tissue in the body. Poly(lactic-co-glycolic acid) (PLGA) nanoparticles containing manganese oxide (MnO) cores have been introduced as novel pH-sensitive contrast agents and can be activated upon tumor cell uptake into low pH endosomes to provide selective MRI contrast at the cancer site through Mn^{2+} generation. The objective of this study was to enhance MRI contrast by reducing MnO core size. Smaller MnO cores should allow for maximum packing volume into PLGA nanoparticles and faster controlled release of Mn^{2+} through a greater surface area to volume ratio. MnO cores were fabricated through thermal decomposition of manganese (II) acetylacetonate using two different temperature ramps (10°C/min and 20°C/min) and aged at 300°C for increasing time periods of 5, 15, and 30 minutes. This study is the first to show that temperature ramping rate and aging time impact both the resulting size and chemistry of MnO cores. Synthesized MnO cores were characterized by transmission electron microscopy (TEM), X-ray diffraction (XRD), and Mn^{2+} controlled release at pH 5 (to mimic endosomal pH inside cells), 6.5 (to mimic the tumor extracellular space), and 7.4 (to mimic the blood). A faster temperature ramp and shorter aging time at 300°C reduced MnO core size to 23.4 nm ± 11.7 nm as shown by TEM. XRD spectra revealed that pure MnO was obtained with longer overall reaction times; however, shorter overall reaction times produced a mixture of MnO and Mn_3O_4. Compared to MnO, Mn_3O_4 is expected to have slower Mn^{2+} production rates. Surprisingly, all synthesized cores had ~46% dissociation to Mn^{2+} over 24 hours in pH 5 media mimicking endosomes. Minimal Mn^{2+} release was found at pH 6.5 and 7.4, indicating that MnO nanoparticles must be targeted intracellularly to obtain sufficient MRI contrast for breast cancer detection. Future studies will investigate the impact of changing the ratio of chemical reactants to achieve further MnO size reduction and decreased polydispersity, as well as develop surface targeting methods to enhance MnO nanoparticle endocytosis within human breast cancer cells.

579939 Data Science and Machine Learning in the Upstream Oil and Gas Industry

Digital technologies have radically transformed many industries in the last two decades, but advances in pervasive data access, machine learning, and digital business models have penetrated the upstream oil and gas industry more slowly. This process is now intensifying, with digital technologies poised to disrupt core business and technical workflows, in production operations, drilling, and subsurface characterization, among other areas. I will review some specific instances where digital technologies, especially those enabled by data science and machine learning, have been successfully utilized within ExxonMobil, drawing lessons for the effective development and deployment of these technologies to achieve transformational business impact.

579937 Big Data Analytics in the Advanced Manufacturing of Biopharmaceuticals

This presentation describes the use of process data analytics and machine learning to improve process productivity, reliability, and control in next-generation biopharmaceuticals manufacturing systems. New information streams are described, including the use of hyperspectral imaging to generate gigabytes of real-time data in the form of high order tensors, and the coupling of on-line automated sampling systems with liquid chromatography-mass spectroscopy (LC-MS) to simultaneously measure transient variations in large numbers of distinct small molecules and proteins during manufacturing. Experimental data are presented from multiple laboratories for a variety of bioprocess unit operations, including bioreactors and freeze-drying systems. The use of that data in first-principles, data-driven, and hybrid models are described, with relationships drawn to chemometrics (e.g., partial least squares, principal component analysis), system identification (e.g., canonical variate analysis), pattern recognition (e.g., Fisher and quadratic discriminant analysis), and machine learning (e.g., ensemble learning, elastic net regularization).

581227 Big Data Impact on the Future of Manufacturing

There is an opportunity for innovation in chemical and pharmaceutical manufacturing processes through the integration of human intuition and experience with big data, machine learning and other software innovations. Subject matter experts bring their experience and expertise for constructing an effective analytics strategy, whereas new capabilities in software enable them to work and analyze processes and outcomes at the speed of thought. This speed of insight is increasingly important in dealing with two of the highest priority pressure points in the industry: time to market from research to production, and replacing batch-based production with a continuous production model.

However, since analysis can suffer from human biases and heuristics, it is imperative to take advantage of big data platforms and applications to ensure completeness of the analytics effort. This includes data collection and aggregation for comprehensive access to production results, as well as advanced analytics to accelerate the time to insight and the breath of questions which may be required for insight. This session will provide a framework for leveraging software innovation in the context of industry requirements with examples of current customer deployments and insights.

580269 Data-Assisted Modeling and Optimization for Process Systems Engineering

Coupling equation-based first-principle modeling with information that comes in the form of input-output data for enhanced decision-making, is gaining significant interest in process systems engineering [1]. Most data-driven optimization techniques consider the system under study entirely as a "black-box", however, in order to take advantage of years of obtained physical process knowledge, this work proposes a hybrid modeling and optimization approach [2-3]. This talk will demonstrate the performance of a suite of flexible modeling and optimization tools that are able to incorporate both physics-based and machine-learning components, to enable reliable data-assisted decision making. Specifically, adaptive and tractable surrogate functions will be developed to represent input-output data, and these will be merged with mechanistic equations in the form of constraints or discrepancy modeling [4]. Optimization is performed via a custom-based branch-and-bound framework that incorporates sampling, machine learning and bounding of the uncertainty of the approximations. Finally, the performance of the proposed methods is demonstrated through a set of process design case studies that rely on multifidelity models and data.

References:

1. Venkatasubramanian, V. (2019), AIChE J., 65: 466-478
2. Boukouvala, F., Hasan, MMF, Floudas, C.A. (2017). JOGO, 67(1-2):3-42
3. Kim, S.H. and Boukouvala, F. (2019), Optimization Letters, 10.1007/s11590-019-01428-7
4. Kennedy, M. C. and O'Hagan, A. (2001), Journal of the Royal Statistical Society, 63: 425-464

575893 Triplet-Triplet Annihilation Upconversion-Based Nanosensors for Fluorescent Detection of Potassium

Ionophore-based optical nanosensors offer many advantages over traditional organic indicators for biological monitoring including response tuning, multi-analyte detection, and ease of development. Typical ionophore-based nanosensors use nile blue-derived indicators called chromoionophores which must contend with strong background absorption, autofluorescence, and scattering in biological samples that limit their usefulness. Here, we demonstrate potassium-selective nanosensors that utilize triplet-triplet annihilation upconversion to minimize potential optical interference in biological media and a pH-sensitive quencher molecule to modulate upconversion intensity in response to changes in analyte concentration. A triplet-triplet annihilation dye pair (platinum (II) octaethylporphyrin and 9,10-diphenylanthracene) was integrated into nanosensors containing an analyte binding ligand (ionophore), charge-balancing additive, and a pH indicator quencher.

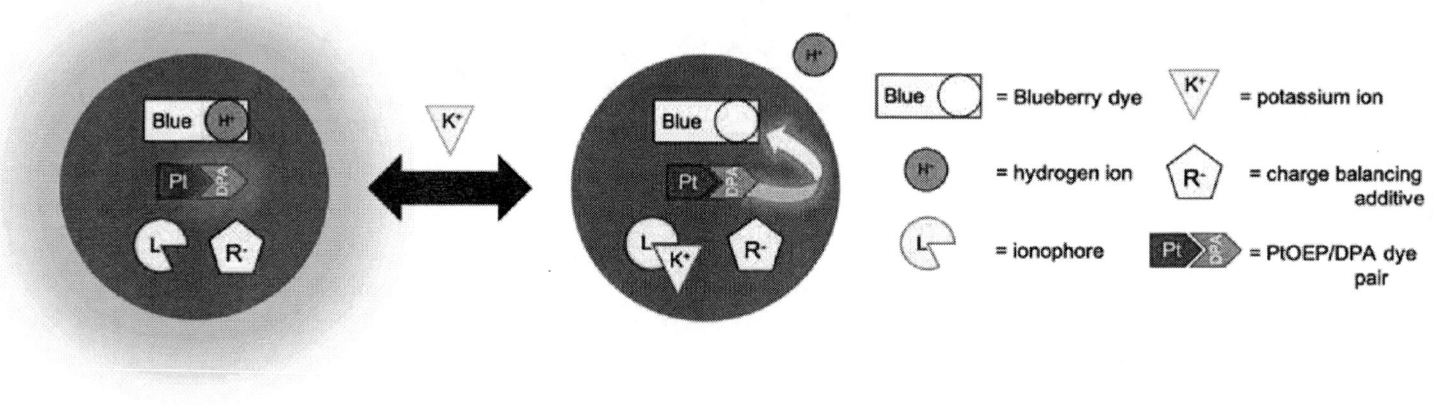

Figure 1. Schematic of nanosensor components co-loaded into the nanosensor matrix and fluorescence quenching mechanism in response to decreasing potassium (K^+) concentrations.

In comparison to lanthanide systems, TTA-UC requires relatively low power excitation and can be wavelength tuned by choice of sensitizer and annihilator. The sensitizer-annihilator dye pairs are organic dyes that are often commercially available, soluble in the organic phase, and able to be implemented into IBOS. TTA-UC has been utilized in several forms as a biological imaging agent. TTA-UC has found limited application in sensing platforms. Thus far, it has been limited to Borisov et al. creating a polymer optode film to directly sense oxygen. However, there are already numerous optical oxygen sensing platforms available. To our knowledge, there have been no reports of TTA-UC used in a more general sensing platform such as those based on ionophores. Nanosensor response to potassium was shown to be reversible and stable for three days. In addition, the nanosensors are selective against sodium and calcium (selectivity coefficient of -0.23 for calcium), the two main interfering ions in biological samples.

578077 Glucose-Responsive Nanoparticles Based on Enzymatic Sensors for Self-Regulated Insulin Delivery

Glucose-responsive nanoparticles based on enzymatic sensors

for self-regulated insulin delivery

Lisa R Volpatti, Morgan A. Matranga, Abel B. Cortinas, Robert Langer, Daniel G. Anderson

Koch Institute & Department of Chemical Engineering, MIT, Cambridge, MA 02139

Motivation: Diabetes mellitus is a disease characterized by poor glycemic control which often leads to severe complications including cardiovascular disease and kidney failure. Many diabetic patients continually monitor their blood sugar and self-administer multiple daily doses of insulin to combat hyperglycemia. Most commercial glucometers rely on the enzymatic sensor, glucose oxidase (GOx), to detect glucose levels. To better mimic native insulin activity, the GOx sensor can also be incorporated into an insulin delivery system. By coupling the sensing with a responsive polymeric material, insulin can be released in a self-regulated manner. Here, we report a glucose-responsive insulin delivery system based on the enzymatic sensing of glucose oxidase that achieves superior glycemic control compared to systems lacking the glucose sensor.

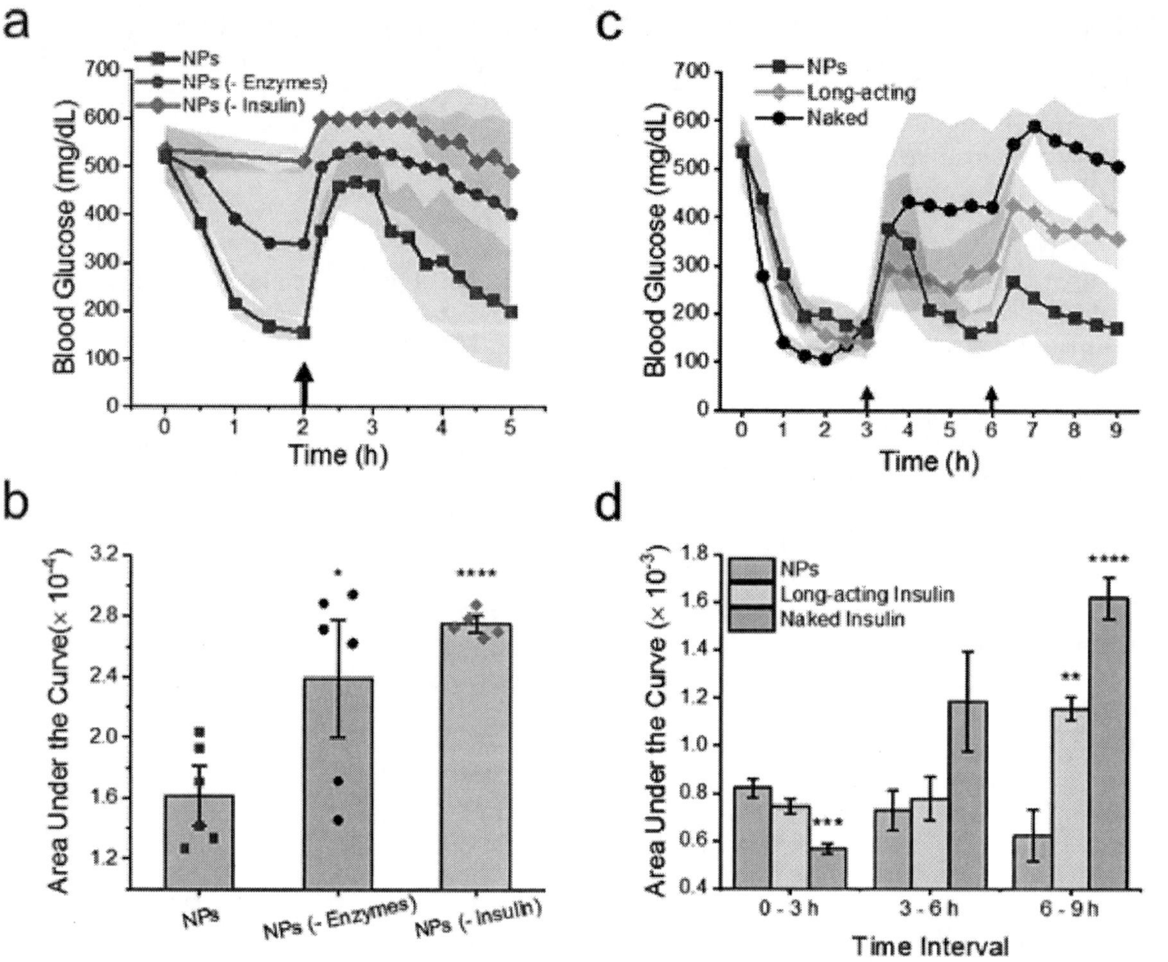

Fig. 1. Glucose-responsiveness provides enhanced glycemic control. a) Response of diabetic mice to a glucose injection (3 g/kg, indicated by arrow) following administration of 9 IU/kg NPs, 9 IU/kg NPs lacking enzymes, or an equivalent dose of NPs lacking insulin. b) Area under the curve for (a). c) Blood glucose levels of diabetic mice following administration of 5 IU/kg NPs, long-acting insulin, or naked insulin with glucose injections at 3 h and 6 h (1.5 g/kg). d) Area under the curve for (c) by 3 h time interval. Shaded regions represent standard deviation. Error bars represent standard error of the mean (N = 4+). Statistical significance is indicated by $^{*}P < 0.05$, $^{**}P < 0.01$, $^{***}P < 0.001$, $^{****}P \leq 0.0001$ in comparison to the first NP bar (blue).

Methods:

GOx acts as a glucose sensor by catalyzing the reaction of glucose to gluconic acid, consuming oxygen and producing hydrogen peroxide as a byproduct. Therefore, polymers that respond to changes in pH, oxidation, or hypoxia may be used in conjunction with this enzyme for self-regulated insulin delivery. Here, acetalated-dextran (Ac-dex) nanoparticles (NPs) were formulated containing insulin, GOx, and catalase. Ac-dex rapidly degrades as a result of the reduced pH from the GOx reaction when glucose levels are high, thus releasing insulin on demand. Catalase provides added biocompatibility by disproportionating the hydrogen peroxide byproduct.

Results: To distinguish between glucose-responsive and non-responsive insulin release, diabetic mice were administered nanoparticles containing enzymes and insulin (NPs), nanoparticles without enzymes (NPs (– Enzymes)), or nanoparticles without insulin (NPs (– Insulin)). Following a glucose injection, only the NP group was able to regain glycemic control within 3 h (Figure 1a). Therefore, the glucose sensor and the insulin are necessary to produce consistent glycemic control with significantly reduced areas under the curve (AUCs, Figure 1b). Diabetic mice were found to have the best glycemic control when administered

NPs, compared to an equivalent dose of a long-acting acylated analog of insulin (commercially known as insulin detemir) and naked insulin (Figure 1c). Specifically, for the time interval between 6 – 9 h, the long-acting insulin and naked insulin groups both have significantly higher AUCs than the NP group (Figure 1d). Therefore, for the same dose of insulin, glucose-responsive NPs provide better extended glycemic control compared to long-acting or naked insulin treatments.

Conclusions: The enzymatic sensor glucose oxidase can provide a trigger for applications in self-regulated insulin delivery. The resulting glucose-responsive nanoparticles provide better glycemic control than a system without such sensing capabilities, which may improve diabetes therapies in the future.

576890 In Vivo Biosensing Platform for Detection of Complex Human Milk Oligosaccharides

Human milk oligosaccharides (HMOs) are prebiotic compounds found in breast milk that favor the growth of beneficial bacteria in the infant gut. While there are many efforts underway to synthesize these beneficial complex carbohydrates, the current standards for their detection are the relatively low-throughput chromatographic methods or mass spectrometry. We present a high-throughput, fluorescence-based detection system for HMOs in vivo in *E. coli* (Enam and Mansell, *Cell Chemical Biology,* 2018). Such high-throughput systems facilitate directed evolution and enables a streamlined synthesis process for prebiotics with a significant reduction in analysis time of new producer strains. We also present combinatorial approaches towards construction of elements for detection of more complex HMOs. Additionally, the reporter enables the use of HMOs as orthogonal inducers of cellular circuits in synthetic biological applications.

564854 Protein Detection with Peptoid-Functionalized Carbon Nanotube Optical Sensors

A limitation to real-time imaging of metabolites and proteins has been the selective detection of biomolecules that have no naturally-occurring or stable molecular recognition counterparts.[1] We present developments in the design of synthetic near-infrared fluorescent nanosensors based on the fluorescence modulation of single-walled carbon nanotubes (SWNT) with select sequences of surface-adsorbed *N*-substituted glycine peptoid polymers.[2] We assess the stability of the peptoid-SWNT nanosensor candidates under variable ionic strengths, solution composition, protease exposure, and cell-compatible conditions, and find that the stability of peptoid-SWNTs depends on the composition and length of the peptoid polymer. We identify a peptoid-SWNT assembly that can detect lectin protein wheat germ agglutinin (WGA) with a sensitivity comparable to the concentration of serum proteins. To demonstrate the retention of nanosensor-bound protein activity, we show that WGA bound to the nanosensor produces an additional fluorescent signal modulation upon exposure to the lectin's target sugars, suggesting the lectin protein remains active and selectively binds its target sugars through ternary molecular recognition interactions relayed to the nanosensor. Additionally, we demonstrate that peptoid-SWNT assemblies present a generic platform for protein detection, by demonstrating extension of peptoid-SWNTs for detection of additional protein analytes. Our results inform design considerations for developing synthetic molecular recognition elements by assembling peptoid polymers on SWNTs, and also demonstrate these assemblies can serve as optical nanosensors for lectin proteins and their target sugars. Together, these data suggest peptoid sequences can be assembled with SWNTs to serve as versatile optical probes to detect proteins and their molecular substrates.

1) Landry, M. P., ... Chio, L., *et al.* Single-molecule detection of protein efflux from microorganisms using fluorescent single-walled carbon nanotube sensor arrays. *Nature Nanotechnology* (2017). doi:10.1038/nnano.2016.284

2) Chio, L., Del Bonis-O'Donnell, J.T., Kline, M.A., Kim, J.H., McFarlane, I.R., Zuckermann, N., Landry, M.P. Electrostatic-Assemblies of Single-Walled Carbon Nanotubes and Sequence-Tunable Peptoid Polymers Detect a Lectin Protein and its Target Sugars. *Nano Letters,* (2019) DOI: 10.1021/acs.nanolett.8b04955

564279 Label-Free QCM Immunobiosensor for Real-Time Detection of GFP Antigen Using IOS-1 Peptoid:

The medical diagnostic industry is in need of cheaper, time-saving, and more robust assays for pathogen detection. Our research approach is focused on the development of short-chain designer peptoids that can be tuned to target detection of specific biological molecules. The need for detection of specific pathogens has necessitated our design of a designer peptoid (IOS-1), which preferentially isolates a target antigen (i.e., pathogen or pathogen metabolite) from a dynamic aqueous stream. When incorporated into diagnostic tools, these designer peptoids, which are customizable to the extracellular matrix (envelope) protein of target pathogens, may proffer a cost-effective and rapid means of pathogen identification and quantification. In this research, the precise Quartz Crystal Microbalance (QCM) is used to track adsorption events in real-time.

In this work, we have developed a technique for the in situ detection of green fluorescent protein (GFP) antigen in flowing PBS buffer solution using an immunobiosensor comprising a gold substrate whose surface is modified ex-situ with IOS-1 peptoid. This peptoid is an oligomer of N-substituted glycines that was designed using the submonomer solid-phase synthesis approach.

The QCM provides real-time frequency shift data for the mounted immunobiosensor during the flow of analyte solution, baseline solution, and reflux solution. By using the Sauerbrey model, mass loading of the adsorbed analyte was calculated, thereby enabling in situ quantification of adsorbed GFP antigen. This immunobiosensor is robust enough for dynamic operation and is reusable in vitro by switching the flow of antigen with 0.1% sodium dodecyl sulfate solution to recondition the immunobiosensor surface before resumed capture of antigen. Equilibrium mass loadings of antigen before and after SDS reflux point toward reproducible surface chemistry for this immunobiosensor. Experimental data obtained using this setup are contrasted with data from a similarly constructed immunosensor consisting of the gold substrate which is surface modified with aminoethane thiol, disuccinimidyl suberate and GFP antibody in a stepwise immobilization scheme.

The average mass loading of peptoid immobilized on the (0.78 cm^2) sensor surface was about 400 nanograms, and this peptoid monolayer can capture between 1 nanogram of the analyte and 200 nanograms of the analyte, depending on the initial concentration of antigen flowed across the immunobiosensor. Characterization of the surface was achieved using FTIR, fluorescence microscopy and XPS. Baseline QCM readings were taken in PBS buffer.

574737 High-Throughput Assaying of Individual Host-Pathogen Dynamics in Influenza A Virus Infection Using Drop-Based Microfluidics

Influenza A virus (IAV) is notorious for causing seasonal epidemics and sporadic worldwide pandemics that have resulted in over 100 million deaths during the past century. Understanding how IAV interacts with its host cells is typically performed using tissue culture systems or with animal models. These systems provide bulk measurements of virus production and/or cellular responses, such as interferon activation and production. However, individual cells exist as a heterogeneous population, resulting in cell-to-cell differences in virus production and antiviral responses. Conventional methods for measuring the antiviral response in individual cells during IAV infection are labor intensive, incur high reagent costs, and require extensive serial dilutions to isolate singularly infected cells. Thus, there is a need for high-throughput analysis of single cells to elucidate individual host-cell dynamics. Here, we have developed a drop-based microfluidic method to characterize IAV infection and quantify the cellular response to viral infection at the single cell level. This method utilizes drop-based microfluidics to encapsulate infected cells within picoliter sized drops followed by single cell, drop-based qRT-PCR analysis. We successfully formed $>10^6$ drops that contained individual human alveolar epithelial (A549) cells infected with the pandemic 2009 H1N1 IAV. Our results show that IAV replicates within the drops and that drops remain stable overnight with high cell viability. We quantified viral RNA and cellular mRNA from these infected drops at kilohertz rates using our novel, multiplexed drop-based qRT-PCR assay. This approach allowed us to measure interferon ß production in response to viral RNA output from $>10^5$ individually infected cells in a single experiment. Our method is the first to measure individual host-pathogen dynamics in single cells following infection with IAV using a high-throughput, multiplexed drop-based assay.

564163 Minimally Invasive Extraction of Plant DNA *Via* a Polymeric Microneedle Patch for on-Site Detection of Plant Pathogens

Molecular diagnoses play an essential role for identification of plant pathogens and global crop protection. However, extraction of DNA from plant cells is challenging due to the presence of surrounding polysaccharide cell walls. As a result, isolation of high-quality plant DNA is mainly confined to well-equipped laboratories, and sample preparation becomes one of the major hurdles to perform molecular diagnostic assays of plant pathogens in the field. To expedite sample preparation and pathogen detection on site, a simple and rapid DNA extraction method has been developed using a polymer microneedle (MN) patch to isolate genomic or pathogenic DNA from plant leaves. This minimally invasive DNA extraction method isolated amplification assay ready DNA from several plant species within a minute. Furthermore, MN extraction of pathogenic DNA was tested and compared to conventional cetyltrimethylammonium bromide (CTAB) extraction by isolating *Phytophthora infestans* DNA from infected tomato leaves for late blight disease detection. For laboratory-inoculated samples, the MN extraction matched the performance of the CTAB method for samples after 3 days of inoculation. For field-collected samples, MN patches successfully isolated *P infestans* DNA from all leaf samples. These MN patches could be used to diagnose other plant pathogens from various plant species, and thus have great potential to become a general sample preparation technique for on-site molecular diagnosis of plant diseases.

576177 Nanostructured Polymer Interfaces for Improved Lectin-Based Capture and Detection of Bacteria

There is critical need for developing materials and interfaces that can be used to capture and isolate bacteria for rapid, culture-free sensing at high sensitivity. Lectin interfaces are considered to be promising for affinity-based microorganism capture and isolation from complex samples, including blood, urine, and wastewater. However, there are limitations associated with the practical use of many lectin-based interfaces due to the fact that the equilibrium dissociation constants (K_D) of lectin-carbohydrate interactions are higher than antibody-antigen binding constants by 2-3 orders of magnitude, resulting in poor cell capture efficiency.

In this work, we have designed surfaces that combine reactive polymer coatings that offer control of biomolecule surface density with nanoscale surface structures that allow for precise control of physical surface topology in order to improve cell capture. First we characterize the critical interface parameters that impact lectin coupling onto flat, reactive azlactone-based block copolymer films that consist of poly(glycidyl methacrylate)-*block*-poly(vinyldimethyl azlactone) ($PGMA_{56}$-*b*-$PVDMA_{175}$). Second, we incorporate well-controlled physical nanostructures into the films using nanospike arrays (black silicon) fabricated from reactive ion etching, which provide underlying physical structures for the polymer coating. The microbe detection sensitivity and bactericidal impact of the both flat and nanopillar polymer surfaces are then characterized.

Our data indicate that the physicochemically-optimized surfaces enhance capture and detection of bacteria from solutions by two orders of magnitude compared to control surfaces, which were lectins immobilized to flat surfaces using standard carbodiimide coupling chemistry (EDC/NHS). We found that flat PVDMA surfaces alone improve the limit of detection (LOD) by one order of due to the fact that the PVDMA polymer offered higher levels of lectin surface density. Addition of the nanopillar arrays to the PVDMA-coated surfaces improved the sensitivity by an additional order of magnitude, down to 10^2 CFU/mL. This is likely due the three-dimensional surface structures that provided enhanced contact with the targeted bacteria. This level of sensitivity gives our interfaces relevance to a variety of bacterial detection applications which include detection in water/wastewater systems (required LOD =10^3-10^5 CFU/mL), urinary tract infections (required LOD = 10^3 CFU/mL), and blood infections (required LOD = 1-100 CFU/mL).

578743 Transcriptional Biosensors That Discriminate between Radiation Sources

Detection of nuclear fuel cycle, enrichment, and weapon development activities is critical for supporting warfighter preparation in chemical, biological, radiological, nuclear, and explosives (CBRNE) operations, nuclear compliance, and clandestine activities. Disadvantages of conventional radiation monitoring and detection systems of being easily identifiable, mandatory placement near a radioactive source for detection, and capacity to report radioactivity at a specific moment in time diminish their applications in the field.In the event of source relocation and contamination, the ability to detect trace amounts of ionizing radiation is a paramount concern for CBRNE operations and prevention of clandestine activities.Microbes in the environment experience behavioral and morphological changes in response to environmental stress, including low-dose ionizing radiation. These responses are known to sometimes persist even after the stress is removed. We hypothesized that these responses can be monitored through transcriptional changes and could be harnessed for biosensors capable of discerning radionuclide type have the potential to monitor and report on nuclear fuel cycle, enrichment, and weapon development activities in diverse environmental conditions.

Transcriptomic profiling through mRNA sequencing analysis revealed unique responses of microorganisms after exposure to acute and chronic radionuclides representative of the nuclear fuel cycle. After quantitative PCR verification, genes with high abundance and overexpression were selected as candidates for biosensor development. Key differences in gene expression were seen in *Pseudomonas putida*, *Escherichia coli*, and *Saccharomyces cerevisiae*after acute and chronic exposure to plutonium-239, tritium, and iron-55 at a dose rate of 8.7 mGy/d. Substantial differences in differentially expressed genes were noted between acute and chronic exposures datasets indicating the potential presence of a radiation-induced time signature. Unique genes that were overexpressed only by a single radionuclide source were selected for toehold switch biosensor development, where selected transcripts trigger fluorescent protein translation from an otherwise repressed state. Future sensor development will include engineering circuits of switch sensors permitting detection of time dependent signals for individual radionuclide sources. In addition to developing passive and autonomous monitoring or clandestine activities, the results of this work will expand upon the limited knowledge of low dose radiation effects in microorganisms.

568538 Competitive Protein-Based Fluorescent Biosensor for Glucose Detection in Exhaled Breath Condensate

Despite recent advances in the development of non-invasive glucose biosensors, reliable measurement of glucose presence at very low concentrations in certain non-invasive fluids is highly desired. The current gold standard for glucose measurements involve electrochemical sensors to detect glucose oxidase (GOx); however, the high dissociation constant of the GOx enzyme prevents it from being able to distinguish sub-micromolar concentration changes of glucose. As alternative non-invasive fluids, such as exhaled breath condensate, are being examined for glucose detection, there is a greater need for a highly sensitive biosensor that can detect and resolve changes in sub-micromolar glucose concentrations. Based on this need, we have developed a fluorescent biosensor using the *E. coli* glucose binding protein (GBP) as a molecular recognition element. GBP is an attractive alternative because the protein has a higher affinity for glucose (K_D = 0.35 μM) compared to that of GOx (K_D = 20 mM).

In this work, we selectively label GBP co-translationally in the *E. coli* recombinant expression system using a chemoenzymatic tagging method. In this method, GBP is labeled with an azide functionality at the N-terminus, which allows for rapid bio-orthogonal conjugation to surfaces using azide-alkyne cycloaddition – a common click chemistry technique. As proof-of-concept, we have demonstrated selective labeling of the GBP to both an alkyne- and dibenzocyclooctyne (DBCO)-labeled fluorophore, TAMRA, directly from *E. coli* cell lysate using copper-mediated and copper-free click chemistry, respectively. Due to their high photostability, large Stoke's shifts, and low pH and temperature dependence, we use CdSe/ZnS quantum dots (QDs) functionalized with DBCO as a fluorescent nanomaterial for bio-orthogonal conjugation to azide-labeled GBP directly from cell lysate[1]. Copper-free click chemistry is employed to avoid irreversible quenching of QDs by copper (Cu^{2+})[2].

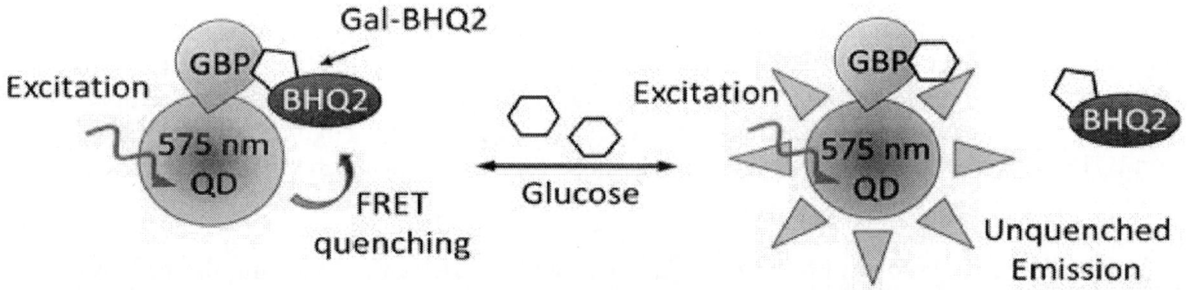

Figure 1. QD-FRET in GBP Glucose Biosensor. Initially, a galactose-quencher (gal-BHQ2) occupies the binding pocket of GBP and results in FRET quenching of the QD. Glucose-containing samples displace gal-BHQ2, resulting in fluorescence emission of QD.

While GBP binds to both glucose and galactose, it has a higher affinity for glucose (K_D = 0.35 μM) over galactose (K_D = 1.4 μM)[3]; thus, galactose can be harnessed as a competitive binding molecule in a fluorescent assay. Förster resonance energy transfer (FRET) is a common method used in fluorescence-based detection, in which the distance between the donor and acceptor molecules determines the rate of energy transfer. The transfer of energy leads to a reduction in the donor's fluorescence intensity and an increase in the acceptor's emission intensity, provided there is close proximity and good spectral overlap between the two molecules.

Here, FRET is utilized to detect the presence of glucose based on the distance between the donor molecule (QD) and an acceptor molecule designed to quench the QD emission. A quencher molecule was designed using a galactose-analog covalently bound to a black-hole quencher (BHQ2) fluorophore that has overlapping emission spectra with the QD. When samples containing glucose are introduced to the QD-BHQ2 system, the galactose-quencher is displaced by the glucose and the fluorescence of the glucose-bound GBP-QD increases proportionally to the concentration of glucose (Fig. 1).

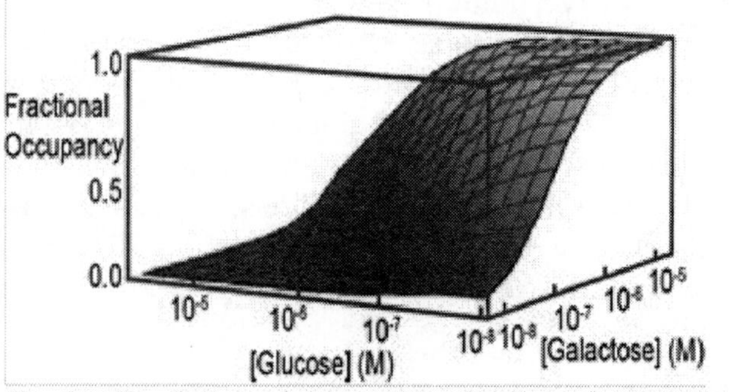

Fig. 2: Simulation of competitive binding of galactose and glucose to GBP. Fractional occupancy of galactose bound GBP is plotted as a function of galactose and glucose concentration 10 sec after glucose is introduced.

A computational model simulating galactose binding to GBP, the introduction of glucose, and the resultant competition of galactose and glucose binding to GBP has been developed using parameters determined by Miller et al. (Fig. 2)[4]. This model indicates that upon the introduction of glucose to galactose-bound GBP, the system arrives at equilibrium within 10 seconds and can provide a linear response over a range of glucose concentrations. As a conservative estimate, at a high galactose concentration (10 μM), a low concentration of glucose (1 μM) can displace over 60% of the galactose-bound GBP. At this level, we would expect at least a 230% increase in fluorescent signal.

This presentation will discuss our work in selectively conjugating the GBP protein onto highly-photostable quantum dots using copper-free click chemistry. Subsequently, we will present the characterization of competitive fluorescent quenching when interfaced with a galactose-quencher on this quantum dot system through fluorescence response curves. To demonstrate specificity of our system to glucose detection at low concentrations, the fluorescence response curves will be compared to those generated with the sensor titrated with galactose, fructose, and other non-specific monosaccharides. Together, this work will enable the detection of low concentrations of glucose such as those found in non-invasive sample matrices such as exhaled breath.

[1] N. Tajarrod, M. K. Rofouei, M. Masteri-Farahani, and R. Zadmard, "A quantum dot-based fluorescence sensor for sensitive and enzymeless detection of creatinine," *Anal. Methods*, vol. 8, no. 30, pp. 5911–5920, 2016.

[2] V. R. Mann, A. S. Powers, D. C. Tilley, J. T. Sack, and B. E. Cohen, "Azide–Alkyne Click Conjugation on Quantum Dots by Selective Copper Coordination," *ACS Nano*, vol. 12, no. 5, pp. 4469–4477, May 2018.

[3] A. V Fonin, O. V Stepanenko, O. I. Povarova, C. A. Volova, E. M. Philippova, G. S. Bublikov, I. M. Kuznetsova, A. P. Demchenko, and K. K. Turoverov, "Spectral characteristics of the mutant form GGBP/H152C of D-glucose/D-galactose-binding protein labeled with fluorescent dye BADAN: influence of external factors," *PeerJ*, vol. 2, p. e275, Nov. 2014.

[4] D. W. Miller and K. A. Dill, "Ligand binding to proteins: the binding landscape model," *Protein Sci.*, vol. 6, no. 10, pp. 2166–2179, Oct. 1997.

569629 Engineering Thermostable Binding Protein rcSso7d Against Zika Virus for Paper-Based Diagnostic Tests

Zika virus is a flavivirus transmitted by mosquitoes and has gained widespread attention in recent years. Although the virus typically only causes mild symptoms in adults, Zika virus has recently been linked to severe birth defects such as microcephaly in infants and other neurological disorders such as Guillain-Barré Syndrome in adults. Therefore, rapid and accurate diagnostics are important to track, control, and prevent spread of Zika virus. As symptoms and diagnostic biomarkers for Zika virus can be similar to other flaviviruses like Dengue virus, it is important that the diagnostic test is specific to Zika virus to minimize cross-reactivity and inaccurate results.

Antibody pairs are widely used as affinity agents in rapid diagnostic tests (RDTs). These tests are frequently designed as a sandwich assay with a pair of antibodies that bind to separate epitopes of the target antigen. One binding protein is immobilized to capture the target disease biomarker, while another binding protein is labeled to produce a signal for detection. Unfortunately, antibodies have some limitations for *in vitro* applications, including difficulty in developing antibodies with adequate stability and minimal cross-reactivity. Therefore, alternate binding proteins have been investigated to replace antibodies in RDTs.

We used the reduced-charged Sso7d protein (rcSso7d) as our protein scaffold due to its intrinsic stability, high yield expression in bacteria, and ease of genetic modification. We previously incorporated rcSso7d binders in paper assays as either the capture binding protein or the reporter binding protein; however, we had yet to demonstrate the use of rcSso7d as both the capture and reporter reagents simultaneously. In this study, we developed a pair of dual-epitope binding rcSso7d binders against a Zika virus antigen, non-structural 1 (ZNS1) protein. We used directed evolution techniques via yeast surface display to engineer clones with binding affinity to ZNS1. Using negative selections, we ensured that the binders would not cross-react with another similar flavivirus NS1 protein, Dengue virus NS1. We incorporated these binders into soluble protein formats as capture and reporter reagents. When the performance of the rcSso7d sandwich was compared to an antibody sandwich in a paper-based format, we found that the rcSso7d sandwich had a lower limit of detection than the antibody sandwich. By demonstrating the use of rcSso7d as both capture and reporter reagents in a diagnostic test format, we validate that rcSso7d is a strong potential antibody alternative in rapid diagnostic tests.

568670 Block Optical DNA Sequencing and Content Scoring for Rapid Genetic Biomarker Identification

A single, inexpensive diagnostic test capable of rapidly identifying a wide range of genetic biomarkers would prove invaluable in precision medicine. Immediate applications of such a technology would be to address the growing threat of antibiotic resistance and in the screening of genetic diseases such as cancer. Current antibiotic resistance diagnostics and profiling assays are often performed only after initial antibiotics fail, and most rely on cell culturing, PCR amplification, and microarray analyses. For cancers or other genetic diseases, diagnostics continue to rely on the identification of proteins, peptides, or gene expression levels. Many of these assays, whether for antibiotic resistance or genetic diseases, are not only time consuming but specific for detecting merely one or a few well-characterized biomarkers. In a push for improved precision diagnostics, we have developed an optical technique for high-throughput, label-free detection of A-G-C-T content in DNA k-mers, called block optical sequencing (BOS) [1]. We demonstrate that these optical measurements of DNA k-mer content can be applied for rapid, broad-spectrum genetic biomarker identification though our robust algorithmic platform called block optical content scoring (BOCS).

The BOS technique uses surface-enhanced Raman spectroscopy for label-free identification of DNA nucleobases from multiplexed three-dimensional plasmonic nanofocusing and characterization of molecular vibrations within the fingerprinting region of ~400-1400 cm^{-1}. While nanometer-scale mode volumes prevent resolution of single nucleobases, our block optical technique can identify relative A-G-C-T content in DNA k-mers (where k ~10 nucleotides). Our robust bioinformatics algorithm, BOCS, uses content-based sequence alignment for probabilistic mapping of k-mer content measurements to gene sequences within a biomarker database, resulting in a probability ranking of genes on a content score. Simulations of the BOCS algorithm reveal high accuracy for identification of single antibiotic resistance genes from pathogens such as methicillin-resistant *Staphylococcus aureus* (MRSA) even in the presence of significant sequencing errors (100% accuracy for no sequencing errors, and > 90% accuracy for sequencing errors at 20%), and at well below full coverage of the genes. Extension of BOCS to cancer and other genetic diseases met or exceeded the results for resistance genes.

These results pave the way for a novel, high-throughput sequencing method with inherent lossy genomic data compression using k-mer identification from multiplexed optical data acquisition. Combined with the BOCS algorithm, this potentiates a test capable of rapid diagnosis and profiling of genetic biomarkers ranging from antibiotic resistance to cancer and other genetic diseases.

[1] Korshoj, Sagar, Hanson, Chowdhury, Otoupal, Chatterjee, Nagpal, *Small* 14 (4), 1703165 (2018).

560751 Rapid and Specific Genotyping for Blood Infection Identification

There is a great need to rapidly identify the organisms in blood infections, and assess their resistance to certain antibiotics, so the correct antibiotic treatment can be implemented from the beginning of therapy. Conventional identification of species and resistance profiling currently takes more than 24 hours.

Our research team has put together a diagnostic process to identify specific bacterial species and the presence of targeted antibiotic resistance genes within 1 hr. The process consists of several unit operations, each of which has been developed individually and is currently being linked into a complete process. The sequence of unit operations follows. First bacteria are separated from blood using a centrifugal sedimentation process requiring only about 2 minutes. Separation efficiency is about 70% even down to 6 CFU/mL blood. Next the bacteria are concentrated by filtration, lysed, and their DNA collected on silica-coated magnetic beads and eluted with an overall recovery efficiency of about 50%. The DNA is reduced in size using sonication. This DNA is run through a capture column having complementary sequences unique to the species and to the antibiotic resistance gene. Sections of captured DNA are then fluorescently labeled with 4 different colors of fluorescing molecules. The bacterial DNA is thermally released from the capture/labeling column in a 3-5 μL volume, which then flows through an optofluidic silicon chip for single molecule detection. Positive fluorescent signal indicates the presence of the specific bacterial species and/or resistance gene. These unit operations have been validated individually for the species *E. coli, K. pneumoniae* and *E. cloacae* and the carbapenam resistance genes *NDM, KPC, IMP* and *VIM*. Linking these unit operations in sequence is much more challenging, and this presentation shows the progress and challenges of building and successfully implementing the entire diagnostic procedure.

574313 Downconversion Luminescent Nanoparticles Harnessing Changes in the Surface Dipole As a Novel Approach for Small Molecule Detection

Optical nanoparticles (NPs) offer a wide range of biosensing applications such as ultrasensitive detection of cancer biomarkers in human serum, multiplexed, point-of-care (POC) detection of analytes, and *in vivo* sensing of reactive oxygen species. While traditional sensing techniques like enzyme-linked immunosorbent assay (ELISA) offers limits of detection (LOD) in the pM-fM range, it is labor intensive, require multiple washing steps, and utilize expensive antibodies for analytical detection which limits their utility for POC applications. Colloidal nanoparticle biosensors can bypass this because they do not require washing and cycling steps and can be designed to provide precise analytical detection without the need for detection antibodies. Optical nanoparticles, such as quantum dots (QDs) exhibit comparable LOD to ELISA coupled with high signal-to-noise ratios (SNR>10). However, the heavy-metal content of QDs results in high cytotoxicity prohibiting the wide-scale use of QDs particularly in clinical applications. Luminescent upconversion NPs have been introduced as biocompatible alternatives with comparable LODs (~6 pM); however, they require expensive, high energy infrared lasers and are limited in their ability for multiplexing. We have identified a new type of inorganic NPs that utilizes downconversion luminescence and changes in surface electric dipoles as a novel approach in detection. The YVO_4 core-shell nanoarchitecture emits a photoluminescent signal originating from the Eu^{3+}-doped core that can be tuned by functionalizing the surface of the nanoparticles with small molecules with weak electric dipoles. We found that the luminescent signal of the NPs can be decreased or increased based on the direction and magnitude of the dipole moment. Similarly, it was discovered that decorating the surface of the NPs with D-biotin generated a positive dipole moment (i.e., electron withdrawing) resulting in a decreased luminescent signal. The surface coverage of the biotin-decorated NPs was evaluated using thermogravimetric analysis (TGA) to perform by time- and concentration-dependent studies into how avidin binding further altered the luminescent properties of the NPs. Exposure to avidin resulted in a further decrease in luminescent signal of the NPs suggesting that the interaction of avidin with biotin resulted in a positive dipole moment with a larger magnitude. This decrease in luminescent signal was found to be concentration-dependent with respect to the free avidin concentration in solution. The downconversion NPs exhibited a similar SNR to upconversion NPs (SNR>10) highlighting their utility as a novel approach in biosensing with the potential to expand to the detection of disease biomarkers or pathogens at low limits of detection.

580445 Nano Field-Effect Transistor-Based Biosensors for Healthcare

Placeholder
One- and two-dimensional nanomaterials are gaining growing interest in biosensors as they provide high sensitivity, multiplexing, small size and portability. Single walled carbon nanotubes (SWNTs), graphene and transition metal dichalcogenide (TMDC) are amongst such nanomaterial that exhibit superior sensing behavior due to their large surface atoms that are highly responsive to surface adsorption events. Combined with their compatibility with modern microfabrication technologies and facile functionalization with molecular recognition elements make them promising candidates for applications in biosensors for healthcare.

In this talk, I will present our recent research on SWNTs, graphene and Molybdenum disulfide nano field-effect transistor-based biosensors for healthcare. Field-effect transistors based on these 1D and 2D nanomaterials were functionalized with biological receptors such as antibodies, aptamers and binding proteins for highly sensitive, selective, rapid, cost-effective and point-of-care monitoring of bacterial and viral pathogens and biomarkers for acute myocardial infarction, psychosocial stress, blood glucose, etc. Further, these biosensors were integrated with paper microfluidics to realize the ASSURED criteria for diagnostics in resource-limited settings.

579640 Transforming Abundant Data into Improved Process Reliability (Keynote Talk)

Today, processing facilities are beginning to effectively utilize the abundance of process and asset health data and information as part of an overall IIoT strategy. Although a digital transformation will require investments in technology, one cannot forget the investment in training staff to behave differently (improved) with the new timely information. It should be noted that some jobs will be replaced (e.g. manual data collection), new roles will emerge (e.g. analysts and planners), and other jobs will evolve to be more efficient. Properly trained operations are invaluable in this process, as they not only act on information from analyzed data, but can also make decisions on their own based on sensor data.

Abnormal operation and imminent asset failure can quickly be identified and corrected with properly trained staff to act on this new timely analyzed information. Big data and IIoT will continue to have a role with shortages of skilled resources, thus having the staff be more efficient and effective at their role with additional information and insight into operations.

This presentation will focus on the applications where big data and IIoT have improved process and asset reliability. Although the applications presented will come from oil refineries, the solutions can be applied to other continuous processing facilities.

579639 Going Digital in Downstream; Refinery of the Future (Keynote Talk)

The last US refinery was built in 1977, to bring these refineries into digital capability is a significant effort and we will explore actual examples and the value that refiners have received. OSIsoft participates in a project to help define the Refinery of the Future. Several examples, like a sensored pump, best uses of IIOT, 3D visualization of the plant and integrity of the asset will be shared to highlight how the downstream business is addressing these technological opportunities. We will share best practices and the progress of recognized leaders in the downstream digital transformation.

579317 Emission Monitoring and Plant Integrity Enhancement

Historically, refineries and chemical plants contribute significantly to emissions of Volatile Organic Compounds (VOCs) and Hazardous Air Pollutants (HAPs). With the increasingly stringent regulations and laws, emission management has to go beyond the traditional Continuous Emissions Monitoring System (CEMS), Leak Detection and Repair (LDAR) approach to prevent potential emission events.

To comply with the new requirement, one prescriptive approach is to enhance plant integrity. Alternative monitoring technologies, such as wireless sensing, optical sensing, facilities the engineers and operators to identify the leaks/emissions conveniently. Information technology, such as industrial 4.0, data analysis and Industrial Internet of Things (IIoT) enables the process industry to communicate and interact effectively. By integrating these technologies, plant integrity, such as predictive analytics, predictive maintenance can be conducted to improve Health, Safety, and Environment (HSE) of process industry.

This presentation describes the relevant environment regulations, emission monitoring practices, and state-of-art emission management technologies, such as predictive flare control and predictive maintenance. Cases of deployments will also be discussed in this presentation.

565604 Optimizing Real-Time Spatiotemporal Sensor Placement for Monitoring Air Pollutant for Health Impact Assessment

Air pollution exposure assessment, which involves the monitoring of pollutant species concentrations in the atmosphere, is an important step in assessing the health impact of air pollution. Currently, the air pollutants are often monitored via stationary monitoring stations. Due to the high cost of maintaining a monitoring network, the monitors can only be installed in limited number of locations. The sparse spatial coverage of monitors can lead to errors in estimating the actual exposure of pollutants in individuals. The recent advancement of cheaper sensors can help address the limitations of using stationary monitors. One approach is dynamic sensing – a new monitoring approach that adjusts the locations of portable sensors in real time to measure the dynamic changes in air quality. The key challenge in dynamic sensing is to develop algorithms to identify the optimal sensor locations in real time in the face of inherent uncertainties in the fate and transport of the air pollutants. The objective of this paper is to develop an algorithmic framework to address the challenge of sensor placement in real time. We demonstrate the capability of our algorithmic framework in a case study in Atlanta. Our real-time sensor placement optimization algorithm will allow, for the first time, the assessment of spatial-temporal variability of pollution which cannot be accomplished by a stationary monitoring station. Besides air pollution exposure assessment, our algorithm can have broader applications in improving sensor placement in drone-based monitoring for water bodies, water security networks, or advanced power systems.

562063 Energy Harvesting Sensors Using Magnetic Phase Transition

Many countries have recently become interested in cyber-physical systems (CPS) in industry and society for sustainable development. CPS equips a data acquisition function, a data storage function, and a network to transfer data. To deploy CPS, it necessary to develop overall security systems, to increase their energy efficiency and to find new energy sources. To overcome these energy-related issues, in this research we investigated the possibility of integrating data acquisition sensors with a recently developed energy harvesting system that combines the magnetic phase transition resulting from changes in temperature, and electromagnetic induction. The proposed system can provide wireless temperature or velocity sensors that directly measure electromotive forces generated by Faraday's law of induction without any additional energy input. Our proposed energy harvesting sensors have the potential to contribute significantly to the development of CPS in the near future.

579901 Carbon Nanotube Photoluminescence Solvatochromism in Biomedicine: Spectroscopy, Imaging, and Modulation

Modulation of the carbon nanotube photoluminescence band position has implications for applications such as sensing and imaging. Carbon nanotube photoluminescence solvatochromism, the hypsochromic or bathochromic shifting of nanotube emission, can be induced via changes in the local dielectric environment. We developed methods to modulate nanotube solvatochromic behavior to improve sensor responses, and we developed tools to observe solvatochromism within live cells and, recently, live animals.

We found that nanotube solvatochromism extends to variations of local electrostatic charge, induced by adsorption/desorption of polyelectrolytes on the nanotube surface, such as proteins and nucleic acids. We also found that dielectric and charge-mediated solvatochromism can be compounded to exacerbate the response. Upon triggered desorption of a polyelectrolyte, such as upon desorption of RNA upon hybridization, for instance, amphiphilic molecules can adsorb to the recently-uncovered nanotube surface. The adsorption displaces water to result in hypsochromic shifting, which can be additive, as polyelectrolyte desorption will also cause a hypsochromic shift. We found that the amphiphile may be a surfactant but, recently, we found that proteins can result in a similar response, facilitating the detection of viruses.

We also developed tools to image changes in nanotube emission in live cells (near-infrared hyperspectral microscopy) and, recently, live animals (preclinical, near-infrared hyperspectral imaging). Carbon nanotube solvatochromism-based sensors, in combination with spectral imaging tools, allow measurements of biomarkers, drugs, and metabolites in live cells and organisms, facilitating improvements in disease detection, drug development, and biomedical research.

575978 Xeno Nucleic Acid Nanosensors for Enhanced Stability

The omnipresence of salts in biofluids creates a pervasive challenge in designing sensors suitable for *in vivo*applications. Fluctuations in ion concentrations have been shown to affect the sensitivity and selectivity of optical sensors based on single-walled carbon nanotubes wrapped with single-stranded DNA (ssDNA-SWCNTs). We herein observe fluorescence wavelength shifting for ssDNA-SWCNT-based optical sensors in the presence of divalent cations at concentrations above 3.5 mM. In contrast, no shifting was observed for concentrations up to 350 mM for sensors bioengineered with increased rigidity using xeno nucleic acids (XNAs). Transient fluorescence measurements reveal distinct optical transitions for ssDNA- and XNA-based wrappings during ion-induced conformation changes, with XNA-based sensors showing increased permanence in conformational and signal stability. This demonstration introduces synthetic biology as a complementary means for enhancing nanotube optoelectronic behaviour, unlocking previously unexplored possibilities for developing nano-bioengineered sensors with augmented capabilities.

570937 Sequence and Aggregation State Determine Stability, Endosomal Processing, and Long-Term Fate of DNA-Wrapped Carbon Nanotubes in Mammalian Cells

Single-walled carbon nanotubes (SWCNTs) have attracted much attention for applications in biomedical imaging and sensing due to their photostable, environmentally responsive near-infrared fluorescence. Single-stranded DNA was found to non-covalently disperse SWCNTs in water while preserving their optical properties and providing a biocompatible wrapping suitable for cellular interactions. The hybridization of DNA-SWCNTs has been studied extensively to determine the relationship between oligonucleotide sequence and SWCNT properties *in situ*, however intracellular environments introduce harsh conditions that can change the identity of the hybrid nanomaterial, altering its intrinsic optical properties. Aggregation via protein-nanotube interactions adds complexity to the biological system, while also causing concerns for cellular toxicity. Here, we show that the stability of DNA-SWCNTs after internalization by mammalian cells is dependent upon the oligonucleotide sequence that wraps the nanotube. The relationship between sequence length, fluorescence stability, and uptake are examined to provide insight on intracellular mechanisms that require considerations in nanotube sensor design. We show that aggregation not only alters the expected rate of uptake into the cells but further affects the endosomal processing pathways. We propose a model for the uptake and expulsion of SWCNTs that accounts for the aggregation state and DNA wrapping. These findings provide fundamental understanding of the interactions between nanotubes and live cells which can be applied towards development of carefully engineered carbon nanotube sensors.

570819 Real-Time Detection of H_2O_2 Signaling in Wild-Type Plants with Nanobionics Approach

As a consequence of their sessile nature, plants have developed complex signaling mechanisms to cope with the highly dynamic environment that they live in. Recent studies revealed that in response to external stimuli, plants can utilize various signaling pathways for cell-to-cell communication to engage different metabolic and genetic machineries involved in the plant defense system. Among the many signaling molecules uncovered, hydrogen peroxide (H_2O_2) represents one of the most versatile signaling molecule in plants in response to environmental stresses such as mechanical wounding. Herein, we develop a nanomaterial-based approach, termed plant nanobionics, using single-walled carbon nanotube (SWNT)-based nanosensors to monitor the wound-induced H_2O_2 signaling pathway in plants. This approach enables non-destructive, real-time detection of changes in endogenous H_2O_2 level in plants at a standoff distance. We demonstrate the versatility of this method to investigate the dynamics between H_2O_2 and electric signal in both model and non-model plants, and show that the nanosensor readout can be paired with portable and inexpensive electronic device. A mathematical model describing the self-propagating mechanism of the wave will be presented. Our work provides a promising tool to elucidate the complex plant signaling pathways across species to aid botany research and precision agricultural applications.

563265 Fluorescent Single-Walled Carbon Nanotubes for Protein Detection

Single-walled carbon nanotubes (SWCNTs) have unique optical and physical properties, and they benefit from the ease of surface functionalization and biocompatibility. Semiconducting SWCNTs fluoresce in the near-infrared (nIR) part of the spectrum, which overlaps with the tissue transparency window. The SWCNTs fluorescence is sensitive to the environment, and depending on the surface functionalization, subtle changes in the proximity of the nanotube can result in significant spectral modulations. Hence, SWCNTs can be utilized as optical sensors enabling real-time optical detection. I will present two recent discoveries of protein detection using SWCNTs functionalized with variants of poly(ethylene glycol). Using high-throughput screening against a panel of human blood proteins, we have discovers sensors for the proteins fibrinogen and insulin. The recognition also occurs in serum environment, showing that the SWCNTs sensors work in this complex environment despite the potential nonspecific adsorption. These results open new avenues for synthetic recognition of biological macromolecules, and hold great promise for medical and clinical applications.

Reference

1. Gili Bisker et. al., ACS Sensors, 3(2), 367-377 (2018).
2. Gili Bisker et. al., Nature Communications, 7 (2016).

563983 Directed Evolution of the Optoelectronic Properties of Synthetic Nanomaterials

Directed evolution is a powerful approach for tailoring protein properties toward new or enhanced functions. This technique allows one to alter a protein in a directed manner even in the absence of a known relationship between protein structure and function. Several nanocomplexes, such as optical sensors based on DNA-wrapped single-walled carbon nanotubes (DNA-SWCNTs), similarly lack a known structure-function relationship. For example, DNA-SWCNT complexes exhibit variations in their fluorescence that depend on the sequence of the DNA wrapping. However, little is known about the relationship between the DNA sequence and its effects on the fluorescence properties of SWCNTs. In this study, we use directed evolution as a guided approach to tuning the optoelectronic properties of DNA-SWCNT complexes through DNA mutation. Applying this technique over two rounds of DNA mutagenesis and screening, we evolved a DNA-SWCNT sensor with a 56% enhancement in fluorescence intensity. Furthermore, this sensor shows retained selectivity and sensitivity towards the analyte of interest. Though we demonstrate this approach to improve sensor brightness, directed evolution offers far-reaching possibilities in the field that can also be applied to improve additional sensor properties such as sensitivity and selectivity.

564858 The Impact of Covalent Functionalization on Single-Walled Carbon Nanotube Sensor Fluorescence and Function

Single-walled carbon nanotubes (SWNTs) have emerged as promising signal transduction elements for biological imaging and sensing due to their notable fluorescence in the near-infrared (NIR) region with minimal light absorbance by water or scattering by tissue.[1] Recently, exciting developments in SWNT surface chemistry have shown the retention and improvement of fluorescence intensity in SWNT after covalent surface functionalization.[2] These chemistries provide chemical handles for specific attachment of molecules of interest to the SWNT surface while keeping the intrinsic NIR fluorescence of the SWNT intact for bioimaging and sensing applications. However, despite their unperturbed fluorescence, it is unclear whether functionalized SWNTs retain their ability for molecular sensing, which necessitates a modulation in SWNT fluorescence provided by the molecular recognition of a surface-adsorbed polymer. We compare the photophysical properties of functionalized SWNTs to their pristine counterparts, and show SWNT optical properties remain available for sensing applications, but only for certain sensing strategies. Molecular recognition provided by surfactant and phospholipid SWNT surface coatings retain their ability to detect molecular analytes such as fibrinogen and insulin. Conversely, DNA oligonucleotide surface coatings under-perform for molecular recognition of dopamine compared to sensors constructed from pristine SWNT. Lastly, we explore the application of functionalized SWNT as dual-functional nanoparticles with both targeting and sensing capabilities. Our work establishes the potential advantages and drawbacks of covalent SWNTs functionalization, despite preservation of SWNT fluorescence, and implications for applications in molecular sensing.

(1) Bonis-O'Donnell, J. T. D.; Page, R. H.; Beyene, A. G.; Tindall, E. G.; McFarlane, I. R.; Landry, M. P. Dual Near-Infrared Two-Photon Microscopy for Deep-Tissue Dopamine Nanosensor Imaging. *Adv. Funct. Mater.* **2017**, *27* (39), 1–10.

(2) Setaro, A.; Adeli, M.; Glaeske, M.; Przyrembel, D.; Bisswanger, T.; Gordeev, G.; Maschietto, F.; Faghani, A.; Paulus, B.; Weinelt, M.; et al. Preserving π-Conjugation in Covalently Functionalized Carbon Nanotubes for Optoelectronic Applications. *Nat. Commun.* **2017**, *8*, 1–7.

566671 Protein Corona Formation on Single-Walled Carbon Nanotubes for Applications in Biological Environments

Engineered nanoparticles such as single-walled carbon nanotubes (SWCNTs) are increasingly used for biological sensing, imaging, and delivery applications owing to their distinctive optical and physical properties. SWCNT-based probes operate at spatiotemporal scales necessary to capture information on chemical signaling such as neurotransmission in the brain[1], and can deliver functional biological cargoes to tissues for genetic tranformations[2]. The critical – and often overlooked – challenge with these nanoscale tools is understanding the fundamental mechanisms of interaction between the nanoprobe and the system they are designed to query. When a nanoparticle enters a biological system, the surface becomes coated with proteins to form the 'protein corona'. Binding of proteins to the nanoparticle not only affects the structure and function of the native biomolecules, but has the added consequences of affecting nanoparticle biodistribution, masking the intended imaging or delivery function of the nanoparticle, and unpredictably re-defining the nanoparticle chemical identity[3].

To apply nanoscale tools in biological systems, it is crucial to understand the mechanisms and outcomes of protein corona formation. We present results on protein corona composition and binding dynamics for our SWCNT-based probes upon exposure to two biological fluids of interest: human blood plasma and cerebrospinal fluid. A selective adsorption assay with characterization by protein mass spectrometry reveals the unique composition of nanoparticle coronae, with identification of key corona proteins. Variation of incubation conditions provides insight on thermodynamic factors and dominant interactions driving protein corona formation. Next, the dynamic exchange of biomolecules in the corona phase is studied by a combination of near-infrared and visible fluorescence monitoring of the SWCNT and fluorophore-labeled proteins, respectively. Ultimately, a 'designer corona' strategy will be implemented to cloak the nanoprobes against untimely clearance pathways, to impart targeting abilities on a cellular level, and to encourage desired biodistribution outcomes on an organism level. Probing the corona composition, driving forces of formation, and timescales upon which the protein corona forms will inform figures of merit for the predictive design and synthesis of nanotechnology-based tools to be applied in protein-rich environments.

References

[1] Beyene, A. G. *et al.* Imaging Striatal Dopamine Release Using a Non-Genetically Encoded Near-Infrared Fluorescent Catecholamine Nanosensor. *bioRxiv* 356543 (2018). doi:10.1101/356543

[2] Demirer, G. S. *et al.* High aspect ratio nanomaterials enable delivery of functional genetic material without DNA integration in mature plants. *Nature Nanotechnology* 1 (2019). doi:10.1038/s41565-019-0382-5

[3] Nel, A. E. *et al.* Understanding biophysicochemical interactions at the nano–bio interface. *Nature Materials* **8**, 543–557 (2009).

578234 Heavy Metal Detection and Biochemical Sensing Using Plant Nanobionics Approach

Plants continuously exchange gases and fluids with the environment above and below ground. Through natural transpiration, they can continually extract water from subsaturated sources and move it across long distances with seemingly no energy input. These properties render plants compelling platforms for the extraction and detection of low concentrations of environmental pollutants in the soil. Several recent studies have attempted to produce genetically engineered plants for phytoremediation and detection of heavy metal contamination. However, such methods are typically limited to a narrow host of plant species amenable to genetic modification. In this work, we develop a nanoparticle-based approach to introduce non-native biochemical sensing capabilities to wild-type living plants. Using carbon nanotube optical sensors embedded in the leaf mesophyll, we demonstrate the detection of heavy metal contaminants through the root uptake of living plants. This approach is non-destructive and can be applied to a wide range of plant species, enabling real-time monitoring of environmental pollutants using plants as self-powered samplers of the environment. We also investigated the sensing mechanism and biocompatibility of the engineered nanosensor. In addition, the biosensing platform can also be used to monitor the endogenous signaling pathways elicited when plants are subjected to environmental stresses such as mechanical wounding. The nanobionic approach creates a new class of biomaterials that can serve as promising detection platforms for environmental remediation and agronomic applications.

576601 Electrochemical Point-of-Care Sensing Methods for the Detection of Tuberculosis Volatile Organic Biomarkers in Patient Breath and Condensate

In this presentation two electrochemical sensing platforms (gas and liquid phase) for the rapid, point-of-care (POC) detection of tuberculosis (TB) will be discussed. According to the World Health Organization, tuberculosis is one of the top ten causes of death worldwide, killing 1.6 million people in 2017 [1]. Up to one third of adults and the majority of young children with TB are unable to produce sufficient sputum for traditional culture testing, making the development of non-sputum-based tests a priority in TB screening and diagnostics. Rapid screening of tuberculosis by evaluation of associated volatile organic biomarkers (VOBs) in breath has been shown to be a promising technology that is significantly faster and more convenient than traditional sputum culture tests. Methyl nicotinate and methyl p-anisate have been isolated as specific biomarkers for the mycobacterium tuberculosis, as they are not found in high concentrations in ambient air or the breath of healthy patients, but should be readily present in all patients with active TB. Previous modelling studies [2] suggest that these biomarkers complex with certain transition metals (such as cobalt and copper) given a particular bias voltage and oxidation state of the metal. This known interaction was used as the basis for the sensing platforms described.

The ability to detect these biomarkers in both liquid (breath condensate) and gas (direct breath) phases is useful in POC settings for several reasons. First, tuberculosis is especially prevalent in low-resource communities, with over 95% of cases presenting in countries classified as low and middle income. Traditional detection methods such as sputum culture have incubation times of days or weeks, and require expensive instrumentation, laboratory level spaces and highly trained technicians. The option of immediately testing collected breath or storing it for testing at a later time in condensate form provides flexibility for both patients and technicians. The sample collection can be performed remotely and non-invasively, making it ideal in high TB burden countries where patients often live far from medical facilities. The need for repeated visits to medical facilities for sample delivery and test results is prohibitive to many, and results in significant patient dropout. It is the hope that these POC sample collection and test methods will increase access to testing worldwide and remove the need for centralized testing facilities. Second, although the levels of TB-specific VOBs are known to be high in extremely sick patients, those at earlier stages of the disease and pediatric patients do not show such high levels. The VOBs described are semi-volatile and readily dissolve in water at relatively high concentrations, making breath condensate a good candidate for simple concentration of the biomarkers from breath. Multiple detection methods for a single patient can also help to increase specificity if the sensors are used in an array. Finally, the nature of the functional metals themselves has been found to be a determining factor in their use in either a liquid- or gas-phase sensor. These features will be discussed further in the overview of each sensing platform.

A solid-state titanium dioxide nanotube array-based sensor is used to directly detect TB biomarkers in the gas phase, making it ideal for POC settings when the breath collected can be immediately tested. Our lab group previously demonstrated the ability of a cobalt-functionalized version of this sensor to specifically detect both of the biomarkers specified using a simple two-electrode setup and a chronoamperometric method [3]. In this configuration, the pure titanium back of the sensor functioned as both the counter and reference electrode, meaning that the applied current is highly dependent on the resistance of the nanotubes and difficult to reproduce from sensor to sensor. To improve reproducibility and specificity of the sensor, the platform has been modified to use a solid-state, conductive electrolyte in order to create a gas phase three-electrode system. A conductive polymer of known resistance is used to separate the functionalized TNA from the silver-painted counter and reference electrodes. Doing this ensures more consistent interactions with the analyte of interest between sensors. In addition to the electrode changes, new scanning potential methods such as cyclic voltammetry (CV) and square wave voltammetry (SWV) have shown promise in detecting the analytes more specifically than chronoamperometry. For example, running cyclic voltammetry using cobalt functionalized sensors shows a reduction peak at about -0.5 V vs a silver quasi-reference electrode when exposed to methyl nicotinate vapor. According to the modelling studies previously referenced, divalent cobalt shows a more specific binding affinity to the biomarkers than copper, and can be easily attached to the sensor surface in hydroxide form using methods such as deposition in an ultrasonic bath or in-situ anodization of the nanotubes. This makes it the primary metal used in this sensing platform.

In the liquid phase, these functional metal interactions can be exploited by mixing the sample solution containing the biomarkers into an electroactive solution (EAS) containing the functional metal ion, and observing the change electrochemically. Although both cobalt and copper can be used in the EAS, copper is preferred when performing liquid electrochemistry because of its highly reversible redox reactions. In contrast, cobalt shows low reversibility in liquid systems, making it more difficult to assess the interaction of cobalt with biomarkers of interest using methods such as cyclic voltammetry (CV). Testing of this platform with a divalent copper EAS has shown that the cyclic voltammogram of the EAS changes significantly when as little as 1 mM methyl nicotinate (MN) is added to the solution. The variation of peak oxidation voltage with MN concentration can be seen in Figure 1 below. In order to further understand the reactions taking place and specifically identify the biomarkers, square wave voltammetry (SWV) is employed to show which oxidation states of the functionalized metal are complexing with the biomarker of interest. The SWV for the copper EAS alone shows three distinct peaks for each of the redox reaction couples present in the voltage window (less than ±1 V). When a small amount of MN is added, the size, shape and in some cases peak-voltage of each distinct peak is affected differently, implying a difference in reaction to specific valence states of copper (Figure 2). When methyl p-anisate (PA) was added at the same concentration, different interactions were observed. In this way, a "fingerprint" method can be used to identify biomarkers once their known interaction is established.

Figure 1: Change in peak oxidation of copper (II)with MN concentration

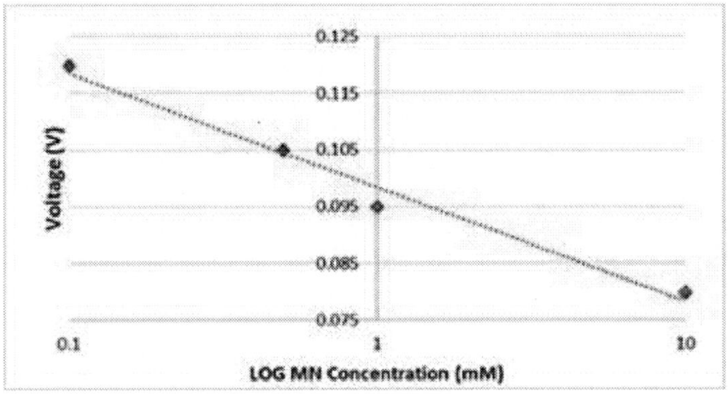

Figure 2: Effect of MN on square wave voltammogram of copper

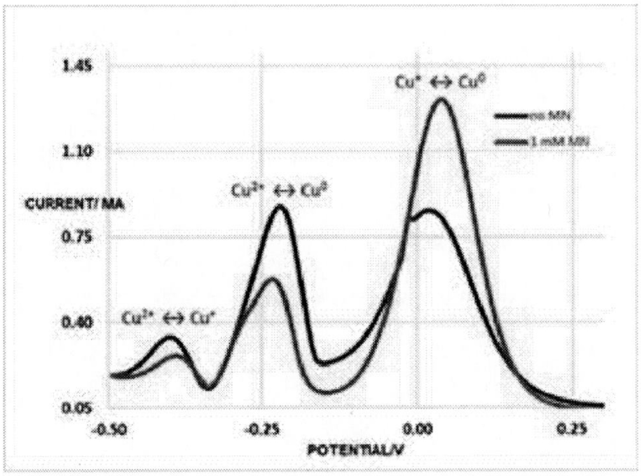

References:

[1] World Health Organization, "Tuberculosis Fact Sheet," 2018. [Online]. Available:

https://www.who.int/en/news-room/fact-sheets/detail/tuberculosis. [Accessed: 12-Apr-2019].

[2] Ray R, Sarma B, Mohanty S, Prisbrey K, Misra M. Assessment of metals in detection of TB biomarkers: Novel computational approach. Journal of Materials Chemistry and Physics. 161 (2015) 1-8

[3] D. Bhattacharyya, Y. R. Smith, S. K. Mohanty, and M. Misra, "Titania Nanotube Array Sensor for Electrochemical Detection of Four Predominate Tuberculosis Volatile Biomarkers," J. Electrochem. Soc., vol. 163, no. 6, pp. B206–B214, 2016.

573972 Reusable Electrochemical Cyclodextrin Biosensor for Resveratrol Detection

In this work, a sensitive α-cyclodextrin (α-CD) based biosensor was designed for detection of resveratrol. Numerous studies have shown that CDs can form inclusion complexes with various organic and inorganic hydrophobic molecules because of their hydrophobic inner cavity and hydrophilic exterior. For that reason, CD-based biosensors have been used for detection of different analytes over the last two decades. However, to the best of our knowledge reusable CD biosensors have not yet been reported, and the current developed CD biosensors are unable to monitor the analyte concentration continuously. During the construction of these biosensors, metal electrodes are mostly coated by CD, without any additional support, and this makes the sensor disposable after each measurement. Therefore, we tried a different approach to develop a reusable CD-based biosensor by combination of self-assembled monolayers and CD supra-molecular interactions. We chose electrochemical impedance spectroscopy (EIS) as our detection technique because of its fast response and high sensitivity.

The technique presented will involve modifying a gold electrode surface with polyethylene glycol (PEG)-SH via thiol chemistry. The surface PEG brushes were then complexed with α–CD. The CD-PEG surface was exposed to different concentrations of resveratrol. We monitored the change in double layer capacitance, which is proportional to the release of CD from the PEG surface. Compared to the bare surface, sensitivity was improved. To check the reusability of the surface, after each serial dilution, the surface was reloaded with CD, and a new serial dilution was performed. It was found that CD was successfully reloaded on the PEG brushes. Attenuated total reflectance – Fourier transform infrared and X-ray photoelectron spectroscopy tests were also carried out to verify impedance results both for initial surface modification and CD reloading. As a control, we did resveratrol serial dilution with a bare surface and a modified PEG surface in which double layer capacitance stayed constant. Moreover, a time control experiment with a CD-PEG surface was carried out. In this experiment we put our modified surface in a PBS solution and measured capacitance every 5 minutes; no changes in double layer capacitance was observed in this experiment either. Finally, the capacitive response of our CD-PEG surface to cortisol was investigated. As it was predicted the α–CD surface only showed some selectivity towards cortisol. Therefore, we have created a semi-selective reusable sensor towards the detection of molecules that interact with α–CD, such as resveratrol in the range 500 pM to 20 nM.

ARTICLE

Received May 2019,

Total iron measurement in human serum with a smartphone

Michael Serhan[a], Mark Sprowls[a], David Jackemeyer[a], Mindy Long[a], Ismael Diez Perez[b], Wolfgang Maret[b], Nongjian Tao,[c] and Erica Forzani *[a]

Accepted 00th January 20xx

DOI: 10.1039/x0xx00000x

Iron deficiency, a leading cause of anemia, is one of the globe's top nutritional disorders according to the World Health Organization. Hemochromatosis, on the other hand, is associated to excess iron, and is usually diagnosed late in the stages of irreversible organ damage. Since abnormally low or high blood iron levels are common worldwide and can be of serious detriment to human health, a ubiquitously available technique for measurement of blood iron could represent a substantial improvement in point-of-care medical technology for monitoring iron-related blood disorders and could potentially spark a trend toward proper early prevention of diseases and health maintenance throughout the life span. Here, we introduce a smartphone-based colorimetric detection system for iron measurement in human serum. The system is designed for point-of-care screening and iron monitoring, and was optimized to be low-cost while still allowing for accurate, rapid iron assessment. It employs a dry sensor strip with optimized chemistry in which iron ions are stripped from blood transport proteins, reduced from Fe(III) to Fe(II), and subsequently chelated with ferene, developing a visible color change for smartphone detection of total iron. We compare the common laboratory iron detection assay of human serum to that of our dry sensor strip. The prototype smartphone assay was sensitive to iron detection with a dynamic range of 50 – 300 μg/dL, sensitivity of 0.00047 a.u/μg/dL and coefficient of variation of 10.5% versus the standard lab approach with sensitivity of 0.00091 a.u/μg/dL and coefficients of variation of 2.2%. Further, a detection limit near 15 μg/dL provided by the smartphone system indicated the system's potential capability for detection of iron deficiencies. ,Finally, drawn human venous blood sample processed for serum and measured for total iron were simultaneously sent to a commercial testing facility (LabCorp) and processed by the smartphone system, averaging errors of less than 3% around the true value of 231 μg/dL. In addition, spectrophotometric validation of the iron detection kinetics for the test conditions were investigated, rendering a more complete insight of the detection reaction. The new mobile-app based colorimetric assay agreed with the standard spectrophotometric method, and provides promising features of mobility and low-cost manufacturing for global healthcare settings.

Introduction

Iron is essential in maintaining health in humans due to reliance on oxygen binding (haeme groups), electron transport (energy production), and as a catalyst of hundreds of enzymes (redox and immune control).[1] Iron metabolism is guided by a complex set of genetically-regulated processes for storage, transportation, and dietary uptake during feeding, thereby providing sufficient iron to all cells regardless of fluctuating dietary quantities, blood losses, or gains via transfusions.[1,2] Both iron deficiency and iron overload can result in numerous physiological dysfunctions and can potentially lead to permanent damage to organ systems such as the liver or the brain.[1,3] Iron deficiency is a frequent concern for menstruating females and in all populations with limited access to proper nutrition, including iron-poor whole foods diets or highly processed/refined foods without fortification.[4] On the other hand, iron-overload can be caused by hemochromatosis a genetic disease (e.g. recessive HFE C282Y allele) with incidence of 1 in 200 people of northern European origin. This disease is characterized initially by vague symptoms such as fatigue that progressively worsen with eventual outcomes potentially including parenchymal damage in various organs and leading to liver disease, pancreatic impairment (diabetes), heart arrhythmias and failure, and neurodegenerative disorders of the brain.[5,6] In fact, over several years, the amount of stored iron in blood can reach 10+ grams (from a normal of 3-4 grams), saturating the transporters, and over-filling or "spilling-over" storage into an increasingly toxic labile intracellular pool.[1,6,7] Thus, iron distribution to and from bodily tissues must be tightly regulated by the human body or with medication to avoid that iron dysregulation eventually leads to loss of function or irreversible damage of organs.

[a] *School for Engineering of Matter, Transport and Energy; Biodesign Institute's Center for Bioelectronics and Biosensors, Arizona State University, 1001 S McAllister Ave., Tempe, AZ.*

[b] *Department of Nutritional Sciences, Kings College Londont, 150 Stamford St., London SE1 9NH, UK.*

[c] *School for Electrical, Computer and Energy Engineering Biodesign Institute's Center for Bioelectronics and Biosensors, Arizona State University, 1001 S McAllister Ave., Tempe, AZ.*

† *Footnotes relating to the title and/or authors should appear here, * Corresponding author: eforzani@asu.edu*

Electronic Supplementary Information (ESI) available: [details of any supplementary information available should be included here]. See DOI: 10.1039/x0xx00000x

Any individual's annual physical exam should include determination of optimal levels of total iron. Unfortunately, due to high cost, only a few proxies for iron metabolism such as hemoglobin and blood red cell count are commonly assessed, but these are poor markers for iron overload, leaving haemochromatosis as a disorder typically detected late in life. Furthermore, regardless of socio-economic status or citizenship, interventions (e.g. supplementation) to prevent or address iron imbalance have no inexpensive or easily accessible measure for efficacy thereby leaving individuals at risk of prolonged state of deficiency or overload.[1,8-10]

Fortunately, total iron together with total iron binding capacity (TIBC) and ferritin (iron storage protein) are clinically validated blood-derived biomarkers that serve as useful measures of iron deficiency or overload.[1,8-10] However, the measurement of all these useful biomarkers is time-consuming, expensive, and painful requiring venous blood draw, temperature-controlled storage and shipping, and use of laboratory-based expertise and instruments such a spectrophotometry.[9,10]

Due to these limitations of clinical assessment of iron metabolism, our goal is to provide a low-cost detection system based on dry, disposable sensor strips to reduce the blood analysis volume by sourcing from capillary puncture, to conduct all analyses at room temperature, and to avoid the need of laboratory space and instrumentation. Thus, we set out to build a point-of-care device for total iron measurement. In this work, we introduce a smartphone colorimetric measurement of total iron from human serum. The choice of utilizing phone-based point of care is due to this emerging technology as a point of care diagnostics. The fact that more than 2.1 billion individuals own at least one smartphone or use smartphones at least once a month, make cell phones the most widespread potential clinical device.[11,12] In addition, recent studies show that since current smartphones can perform complex imaging algorithms and are widely prevalent, they can massively be adopted for point of care and clinical applications which include colorimetric, fluorescence and spectroscopic analysis.[13,14,15,16]

We have chosen to initially focus on total iron since iron is: i) the mineral of interest in preventing disorders related to iron metabolism and ii) is also measured in TIBC when determining transferrin % saturation. Transferrin is the protein responsible for transport of iron in blood, and the necessary additional steps for TIBC measurement have been examined and considered in current design, but the results and integration will be the subject of forthcoming publications. In addition, the separation of serum/plasma from whole blood will be described as a separate publication subject also. Here we focus on development of a dry reagent sensing strip for iron detection in serum samples that gives accurate point-of-care performance, is easy to use (handling and disposal), and has reasonable shelf-life at room temperature, and low manufacturing cost.

Experimental Methods

Reference method for iron detection

A reliable certified laboratory reference method for iron quantification by spectrophotometric assay detects peak absorbance values in the 575-600 nm range for a ferrous complex with ferene (Fe(II)-ferene). The certified protocol begins with whole blood; samples are processed to isolate serum or heparinized plasma, which is then processed with citric acid to strip Fe (III) from transferrin , followed by reducing Fe(III) to Fe(II) with ascorbic acid, and finally the chelation of Fe(II) to the chromophore ferene.[17,18] Ferene is chosen because of the direct proportionality of iron concentrations to absorbance values from the ferrous complex.[17-18] In addition, thiourea is used to quench Cu(II) (a known interferent in blood iron detection) with a protein denaturant (surfactant), sodium dodecyl sulphate, for reaction homogeneity. The incubation time for the final reaction is five to thirty minutes.[17]

Two sources of iron were used throughout our study for all measurements (in triplicate): 1) a set of iron standards and 2) human venous blood draws, processed for serum, from the latter of which a portion was sent to LabCorp for measurement and thus external validation.

Optimized method for iron detection in serum

The reference method introduces 200 mM citric acid, 34 mM ascorbic acid, 100 mM thiourea, and surfactant ("reagent A") to the standard or serum sample, followed by ferene at >3 mM ("reagent B"), with final volume ratios of reagent A : reagent B : sample of 5:1:1.[17] Two issues with the *reference method* encouraged us to study and optimize the method: 1) an interest to increase sensitivity and 2) a need to avoid protein precipitation during the incubation of serum samples. Increasing iron-containing sample and ferene concentrations by simply reducing the "reagent A" volume toward reagent A : reagent B : sample ratio 3:1:1 proved to be fruitful (see Results), leading to an *"optimized method"*. **Table 1** shows the resulting final molar ratios of the "reference" and "optimized" methods right before using the spectrophotometer. More information on the rationale behind sensor strip ratios, is given in the section *sensor strip design*.

Table 1: Comparison of molar concentration ratios of reagents to iron for different analytical methods used in this work.

Reagent	Reference Method	Optimized Method	Sensor Strip Method
Ascorbic acid	9.5×10^3	5.7×10^3	2.7×10^3
Citric acid	55×10^3	34×10^3	16×10^3
Thiourea	28×10^3	17×10^3	8.0×10^3
Ferene	2.2×10^2	2.2×10^2	3.2×10^2
Iron	1.0	1.0	1.0

Note: Ascorbic acid, citric acid, and thiourea are far in excess (2,700-55,000 times larger than the iron concentration), whereas the ferene is in excess rendering ~220-320 ratios. The calculations are done for practical examples where 50 µL of iron standard (100 µg/dL) was added to 200 µL of total assay reagents in the optimized method, 36 µL of the same iron standard was added to 214 µL of total assay reagents in the reference method, whereas 30 µL of 100 µg/dL iron sample was delivered to the dry sensing channel in the sensor strip method.

The surfactant, sodium dodecyl sulphate precipitates proteins in serum samples, thereby causing unusually high absorbance values due to sample turbidity. Thus, we completely removed the surfactant from the *reference method* for the *optimized method*, in addition to decreasing the concentrations of "reagent A" (as explained above).

Spectrophotometric determination

Except when noted otherwise, all spectrophotometric measurements were performed with 96-well plate in a Spectra Max M5 spectrophotometer at 590 nm, using 250 µl final test volumes, which rendered a path length of 0.6 cm. Iron standards were made fresh from Fe(III) nitrate nonahydrate crystals in 0.5 M nitric acid solution, by using high-intensity shakers for 15 minutes to ensure iron crystals were completely dissolved. Calibration curves were obtained from analysis of a blank and 25, 50, 100, 150 and 300 µg/dL iron standards, which covered the range of physiologically relevant total iron levels. Both *reference and optimized methods,* plus all dry sensor strip experiments were conducted by the same technician, on the same instrumentation over the course of several months. In addition, 20 whole blood samples were obtained via consent of a subject, using Arizona State University's IRB study protocol (STUDY00008255). The samples were processed from human venous blood to serum and were measured and compared against a simultaneous measurement at LabCorp. In addition, reaction rates of the reaction were recorded and analysed.

Specificity of the iron detection reaction

In order to study the selectivity of the iron detection reaction, we tested the response to several potential serum interferent analytes, including 107 mg/dL urea, 1,332 mg/dL sodium chloride, 37 mg/dL potassium chloride, 6.8 mg/dL uric acid, 5 mg/dL creatinine, and 26 mg/dL glucose, and used the *optimized method.*

Kinetic investigation of the optimized method

In order to develop a better knowledge of the iron detection reaction under the conditions of our *optimized method,* we studied the reaction kinetics, using 50 and 100 µg/dL iron standards in combination with 2 and 4 mM ferene concentrations. We used time profile of absorbances and their corresponding numerical derivatives to determine the reaction order and rate constants. This analysis enabled the rational selection of the iron detection reaction times, which were critical for the development of iron detection on the dry sensor strips.

Sensor strip design

We focused on iron analysis from serum with the goal of developing and optimizing a dry reagent sensor strip for accurate, sensitive, and reproducible detection of iron. We build the sensor strip based on a simple vertical flow design with the aim of iron detection time of five minutes. The sensor strip has a sensing channel and a reference channel (**Figure 1**). The sensing channel has a layer made of a dry, porous and absorbent nitrocellulose blotting filter paper impregnated with all reagents and a build-in capacity to drive a 30 µL sample by capillary forces without spilling, and rapidly separating large components such as proteins from soluble ions.[22] This sensing layer contains reagent A, which strips Fe(III) from transferrin, reduces Fe(III) to Fe(II), and chelates potential interferents such as Cu(II). The sensing layer is also impregnated with reagent B Ferene to chelate Fe(II) to form the coloured complex. The reference channel is nearly identical to the sensing channel, but without any chemical modification. Thus, the sensing layers separate the sample, contain the chemical reactions, and produce a color change, whereas the reference material does not accept sample and is used to measure reference lighting conditions calibration. The color change is read via reflection of light resulting from white LED incident light on the sensor surface. The sensor housings were designed in SolidWorks, and 3D printed using an Ultimaker 3 printer, and Polylactic acid (PLA) feed polymer.

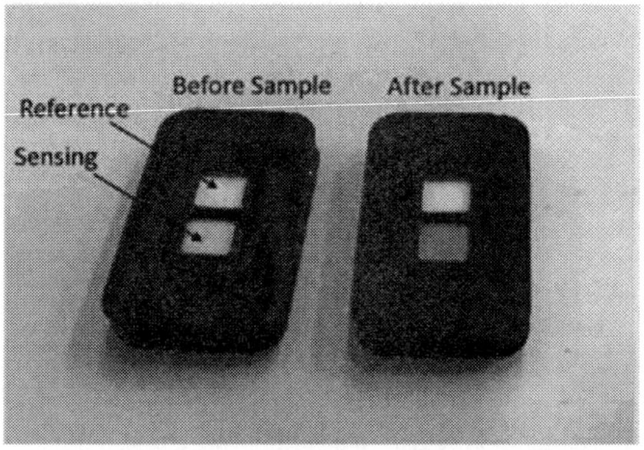

Figure 1: Sensor strip sensing and reference channels before and 5 min after delivering a 30 µL sample.

In order to estimate the final reagent concentration ratios in the dry sensor strip, we weighed 14 dried sensing channel blotting filter papers prior and after the impregnation with the reagents, and calculated weight differences to assess the total weight of reagents. Assuming the reagents (ascorbic acid, citric acid, thiourea, and ferene) were absorbed maintaining their solution mass concentration ratio, we estimated the final reagent concentration once rehydration of reagents occur with the 30 µL sample (**Table 1**).

Sensor strip stability to heat exposure

In order to explore the stability of the sensor to heat exposure, we performed accelerated tests by placing the sensors in an oven at 50 °C for 2 days. Based on predicted accelerated test algorithm, the accelerated aging factor (AAF) defined as the ratio of the room temperature estimated time and the accelerated aging time was as follows:

$$AAF = 2^X \quad (1)$$

with

$$X = \frac{T_{heat} - T_{room}}{10} \quad (2)$$

Sensors were only heated for 2 days at a temperature difference of 30 Celsius. These conditions represent 16 days of accelerated aging with no loss in sensitivity.

3D-printed design of chamber, smartphone mount, and sensor strips

~~A~~ To secure the new sensor strip and to control the smartphone camera's position while ensuring proper lighting, SolidWorks and Ultimaker 3 were used to build a sensing chamber to house the strips for a mounted smartphone. As seen in **Figure 2**, an appropriate focal distance from the sensor strip was required for the Apple smartphone used in our experimentation. A light source was mounted inside the chamber, requiring a coin cell battery powering white LEDs with a difusser.

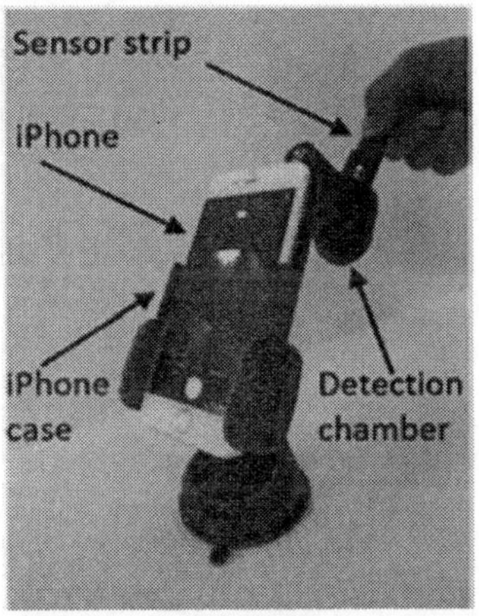

Figure 2: Apple smartphone secured to 3D printed case that integrates the sensing chamber with an LED lighting, and a specific app for analysing iron in serum dripped into a single, secured sensing strip. The image shows the user inserting the sensor strip into the chamber.

All materials were inexpensive or 3D-printable components thereby promoting low-cost equipment designed around smartphone imaging and simple blood processing. See table 2 for a detailed cost breakdown, with a device "capital" cost of $13 and sensor strip components costing less than $0.07 each measurment, in comparison to Lab-based iron measurements of $25/each. Again, each use of the device requires replacement of sensor strip and serum preparation of the venus sample ($0.6 per sample).

Further, the sensor strip is under improvement to accommodate whole blood from human capillary puncture.

Table 2:cost breakdown of device

Device Cost	Cost ($)
3D-printed chamber	2
LED circuit	8
Android charging cable	3
Replacement Sensor Strip Cost	**Cost ($)**
plastic	0.04
membranes	0.0024
reagents	0.017
Additional costs	**Cost ($)**
pipette tips (per measurement)	0.0082
Lithium heparin tubes (per measurement)	0.05
syringue with needles (per measurement)	0.5
centrifuge (one time cost)	50
200 uL pipette (one time cost)	50

Additional fixed costs include a centrifuge (~ $50) and a 200 uL pipette (~$50). With this is mind, these sensors can be deployed in different regions of the world and iron assesments can be conducted on a point of care level.

Smartphone measurement of iron on the sensing strip

Apple provides high quality imaging hardware and software and therefore, it was our choice for colorimetric detection of iron. It boasted an 8-megapixel iSight camera with 1.5µ pixels, autofocus, ƒ/2.2 aperture, hybrid IR filter, exposure control, and with the flash turned OFF. Colour changes on the sensing channel and values from the reference area were electronically captured and analysed by an in-house iOS App developed for this purpose, which deconvolutes the image's Red, Green, and Blue (RGB) components. The operational performance of our in-house iOS App was compared with the publicly available ImageJ software which requires a windows or mac operating device and separate camera to determine if the smartphone could process colours as quickly, capture high-enough quality images, and be easily programmed for a user-friendly experience. First, a wide variety of colours were printed on paper by an RGB generator, systematically analysed by both technologies, and the resulting RGB intensities compared for agreement (see supporting information section). Second, using the iron standards (25 µg/dL to 300 µg/dL) on our sensor strips, sensing and reference channels were read by both technologies and absorbance signals were calculated using the following equation:

$$\boldsymbol{Absolute\ absorbance} = -\boldsymbol{log}\left(\frac{\boldsymbol{I}_{\boldsymbol{sensing\ channel}}}{\boldsymbol{I}_{\boldsymbol{reference\ channel}}}\right) \quad (3)$$

where I = intensity rendered from RGB component deconvolution of the colours.

Statistical analysis

In the effort to assess accuracy and determine variability in the measurements of the smartphone when compared to the reference and modified method for iron detection, we chose the dimensionless statistic coefficient of variation (C.V (%)):

$$C.V(\%) = \frac{\textbf{standard deviation of slope}}{\textbf{average of slope}} \quad (4)$$

where "slope" represents the sensitivity (Absorbance change vs. known iron concentration).

Results and discussion

Optimized method for iron measurement

In order to confirm the reported maximum absorption wavelength of the iron complex, spectral analysis of the Fe(II) – ferene complex was performed. A single sharp peak with a maximum absorbance at 590 nm was consistent with literature[11-13] , and it was observed throughout the different applied methods: reference method and optimized method. Calibration curves based on the absorbance changes at 590 nm as a function of the iron standard concentrations were made and presented a linear dependent signal as a function of the iron ion concentration . The sensitivity of the optimized and reference method was 0.00093 a.u./µg/dL, and 0.00072 a.u./µg/dL), respectively with a regression coefficient of 0.99 for both methods (**Figure 3**). In addition, the CV of the optimized and reference methods was 2.2%, and 3.7%, respectively. Thus, the sensitivity of the optimized method was 30% greater than the corresponding value of the reference method, and the CV of the optimized method was 40% smaller than the one from reference method, giving a higher quality standard method by which to validate the results of smartphone analysis of total iron in human serum samples.

Measuring the effect of protein precipitation

Serum samples analysed with the reference method showed irreproducible and highly variable absorbance changes at both early in the incubation as well as throughout 60 minutes. **Figure 4** shows a profile of iron detection absorbance change at 590 nm vs time for a serum sample of 231 µg/dL total iron, using the reference method and the optimized method. The iron complex absorbance for the reference method (circles) showed at 10 minutes a value that was four times higher than the absorbance from the optimized method (squares). The reference method's values were equivalent to a concentration output reading of nearly 1,000 µg/dL. On the contrary, the optimized method (with zero detergent) produced reasonable results within 3% error of 231 µg/dL iron concentration readout. Furthermore, the iron complex absorbance readings (circles) for the reference method remained still three times higher than those for the optimized method even in following incubation times of 60 minutes, which is the suggested incubation time for reference procedures and processing of serum or plasma samples. **Figure 4** also shows the absorbance of the iron detection reaction at a non-absorbing wavelength of 730 nm for the reference method (triangles). The non-zero absorbance values and the significant changes of absorbance were indicative of turbidity. This led to the elimination of detergent for both, the optimized reference method as well as for the sensor strip design. It is important to mention, that the elimination of the detergent enabled to shorten the iron detection time to 2 minutes. This is a very advantageous feature that it is necessary when designing a new point-of-care analysis methodology.

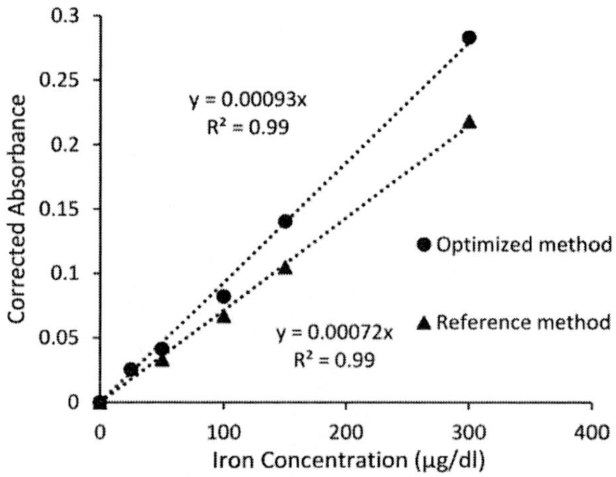

Figure 3. Calibration curves compared. The reference spectrophotometric method consists of a 5:1:1 volume ratio (reagents A : reagent B : iron standards) and gives a slope of 0.00072, whereas our optimized method with 3:1:1 volume ratio increases sensitivity by 30%.

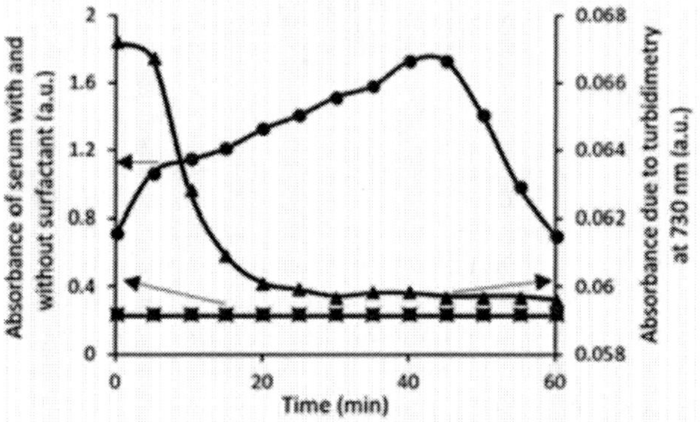

Figure 4. Left y-axis: Profile of iron detection absorbance change at 590 nm vs time for a serum sample of 231 µg/dL total iron, using the reference method (circles (•)) and the optimized method (squares (■)). Right y-axis:

Turbidimetry analysis - profile of absorbance change at an iron complex non-absorbing wavelength (730 nm) vs time for the same serum sample of 231 µg/dL total iron, using the optimized method (triangles (▲)).

Specificity of the optimized method

The optimized method's iron detection response to 50 µg/dL of iron at 590 nm was compared to potential interferents. **Figure 5** indicates that the interfering signal from the combination of urea, sodium chloride, potassium chloride, uric acid, creatinine, and glucose was negligible .

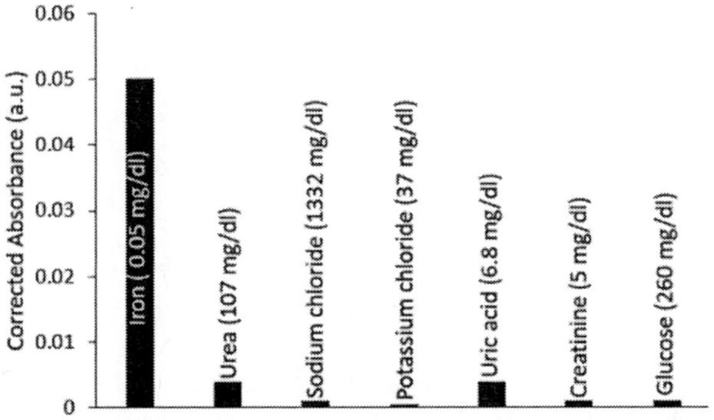

Figure 5. Specificity test, using the optimized method's iron detection reaction. Comparison of the corrected absorbance values of the analyte of choice (R for iron) versus urea (1), sodium chloride (2), potassium chloride (3), uric acid (4), creatinine (5), and glucose (6). Further, the combination of these six interferences do not contribute to measurement error.

Finally, 20 human serum samples were read using the optimized method with 2 minutes incubation time. The average of 238 µg/dL (7.5% CV) came to be 3% (seven units) higher than the LabCorp value of 231 µg/dL. This result demonstrated the accuracy level of the optimized method's iron detection reaction in comparison with the standard-of-care method applied in clinical central laboratories.

Kinetics results using optimized method

In order to gather more insights of the iron detection reaction, spectrophotometric kinetic studies were performed. We initially determined the molar extinction coefficient of the Fe(II) -ferene complex (COM) under the 'optimized method' conditions. The resulting extinction coefficient was 33,366 $L.cm^{-1}.mol^{-1}$, utilizing the slope of the calibration curve for the optimized method, a path length of 0.6 cm, and the Beer-Lambert law. The assessed value was similar to that reported in literature (34,500 $L.cm^{-1}.mol$).[23,24] The rate of the Fe(II) - complex (COM) formation in presence of ascorbic acid (AA), oxidizing to dehydroascorbic acid (DAA) (*k*) is as follows:

$$2Fe(III) + H\text{-}AA\text{-}H + 6ferene \xrightarrow{k} 2[Fe(II) - 3ferene] + AA + H_2$$

However, considering the fact that the AA concentration is highly in excess, the overall reaction rate (k') can be simplify to:

$$Fe(III) + 3ferene \xrightarrow{k'} Fe(II) - 3ferene$$

AA in excess

Under the above-described conditions, the reaction order for iron (α), and ferene (β) could be determined according to the following rate law:

$$\frac{d[COM]}{dt} = k'[Fe^{3+}]^{\alpha}[ferene]^{\beta}$$

Details are shown in the Supplementary information and summarized in **Table 2**, which indicated almost a first order reaction for iron ion, and almost a second order reaction order for ferene.

Table 2. Kinetic Parameters of reaction rates (*k*, *k'*) and reaction order for iron (α), and ferene (β) from studies of the ferrous complex formation in presence of ascorbic acid (AA) in excess.

Parameter	Value
Alpha (α)	1.2
Beta (β)	2.3
k' ($mM^{-1.47}s^{-1}$)	0.089
k ($mM^{2.47}$ s^{-1})	0.0044

Smartphone and dry sensing strip

First, we tested operational performance of the in-house built iPhone App using ImageJ software as a reference method. Linear relationships of 1.01, 1.06, 0.90 with null y-axis intercepts (and regression coefficients of 0.98, 0.99, 0.97, respectively) for the Red, Green, and Blue component signal correlations between the two methods were obtained (see Supplementary Information section, *Smartphone app validation*). The results indicated that the App was accurately assessing the RGB component signals from the images.

Second, iron standards were applied to the dry sensor strips. As example, **Figure 6** shows screenshots of the iPhone App for the sensor strip during the image capture (a), and result output (b) while **Figure 7** shows a close up feature of the sensing channel of different sensor strips exposed to increasing iron standard concentrations, and serum. By following analysis of each sensor strip after the exposure to increasing iron ion standard concentration solutions, a calibration curve was obtained with consistent and reproducible features. Both smartphone App and ImageJ's results were compared, and resulted in equivalent calibration curves, which further validated the design and software deconvolution algorithm of the in-house built App. Overall, the Red component was shown to be most sensitive compared to Green and Blue components, producing a slope of 0.00049, $r^2 = 0.96$, followed by Green (0.00032, $r^2 = 0.97$). On the contrary Blue components did not show sensitivity to the iron concentrations. Red was thus chosen as the sensing signal for the smartphone (**Figure 8**). Overall, the smartphone quickly captured high-quality images for sufficient processing and was programmed for a user-friendly experience.

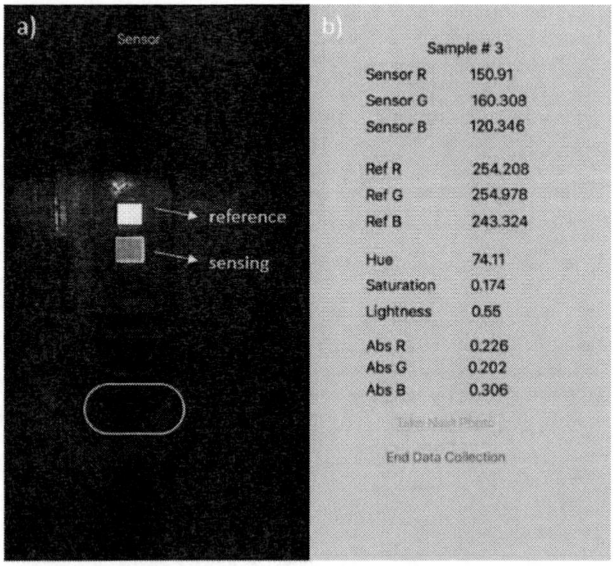

Figure 6: Smartphone application user interface showing a) camera's view of the sensing and reference channels on the dry sensor strip and b) iOS app output screenshot after sensor strip image is taken.

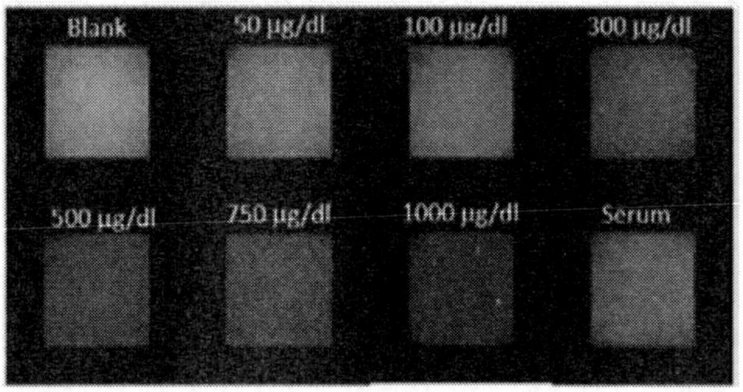

Figure 7: Smartphone application's images of the sensor strip's sensing area upon exposure to increasing iron ion standard concentrations and to serum.

Next, 30 calibration curves of 25 to 300 µg/dL were repeatedly analysed for consistency with new sensor strips prepared from different batches and the smartphone method. The calibrations showed a linear relationship between the Red absorbance signal and the iron concentration, with a slope of 0.00049 and a coefficient of variation of 10.5% (**Figure 8**). In addition, an estimated detection limit (LoD) of 16.5 µg/dL total iron concentration was calculated from the assessed sensitivity, and by assuming a signal equal to 3 times the noise level from 30 blank samples (see LoD indicated by a red asterisk in **Figure 8**).[25,26] Using calibration curves from this new method and the optimized method, serum samples were processed and measured within 2 minutes. The analysis of multiple serum samples from the same subject by the optimized method gave an average of 238 µg/dL (7.5% CV), coming 3% (seven units) higher than the LabCorp value of 231 µg/dL. On the other hand, the new smartphone method gave an average of 227 µg/dL with a 4% C.V, coming even closer. Despite the small differences in results, the overall accuracy of the new smartphone-based proposed method was greater than 90%, which is acceptable considering the simplicity of the system and order of operations and procedure of practice.

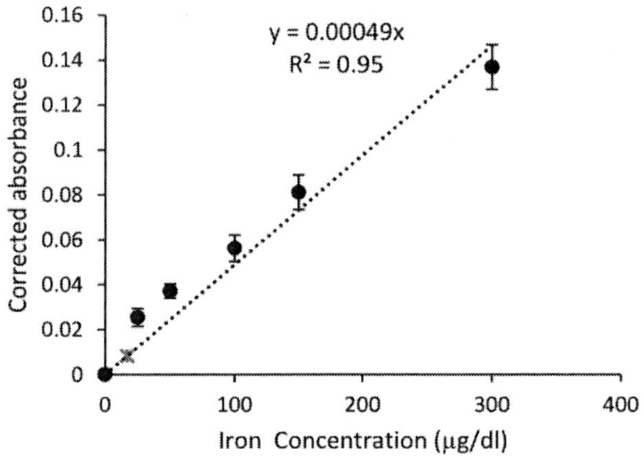

Figure 8: Smartphone calibration curve from 25-300 µg/dL gives a reasonably high slope (compared to the optimized method of 0.00093). Data averages are marked with a +/- single standard deviation range. Further, the estimated limit of detection is indicated by a red asterisk.

Overall, the calibration curve sensitivity of the smartphone, as well as with the accuracy of iron ion detection of initial human serum measurement were satisfactory. The performance characteristic features of the new smartphone-based method and associated device can be used for screening and monitoring of iron, with the main goal of identifying iron deficiencies, overload and normal ranges. In fact, of biological and clinical interest, the elevated total iron value of our subject's case study indicated early hemochromatosis. Shortly after, a diagnosis of the disease was corroborated by analysis of the subject's ferritin (212 ng/mL) and transferrin saturation (93%+), as well as genetic testing that confirmed homozygous C282Y alleles for the HFE gene, using *23andme* services.

Table 3 summarizes the sensitivities of the new smartphone-based method and the modified method, as well as the CV associated to the sensitivity slopes from multiple experiments, and the multiple time analysis of a single serum sample.

Table 3. Summary of sensitivity of optimized and smartphone methods and analysis of serum sample*.

Reading Method	Slope	C.V (%)	Error (%)	Number of replicates
Smartphone				
Calibration Curve	4.9×10^{-4}	10.5	-	8

Serum validation*	-	4.0	2	3
Spectrophotometry				
Calibration Curve	9.3×10^{-4}	2.2	-	24
Serum validation	-	7.5	3	19

A LabCorp analysis gave a result of 231 μg/dL for Total Iron in whole blood, whereas our smartphone-based method gave 227 μg/dL and our optimized method gave 238 μg/dL. Error (%) indicates simply the difference between the experiment and the LabCorp analysis.

Conclusions

Inspired by successful glucose control for management of diseases (Type I and Type II diabetes), and by former experiences of implementing technologies connected to mobile devices[27-29], we have completed several steps toward creating a point-of-care iron ion measurem___ent scheme for prevention or intervention in human iron deficiency or toxicity. First, we have created an internal reference system that can be used to validate iron detection; second, we have tested the chemistry for serum processing in an inexpensive dry sensor strip; third, in-house built application (App) development made available by Apple (iOS) enabled our rapid reading of sensor strip chemistry for testing accuracy and reproducibility; fourth, software and 3-D printing technology empowered our design and testing of a measurement chamber; and last, our methodical approach to optimizing chemistry and colour-change processing has led to a look into the kinetics involved. Our experimental data confirms a linear relationship between iron ion content and absorbance in the physiological ranges for serum iron ion levels, and this combined to the five major accomplishments naturally encourages us to proceed with capillary blood processing and integration of TIBC. This next phase will require cellular material to be filtered (suspended or bound) to the top matrix layer, leaving transferrin-containing serum to complete the relevant chemical reactions, ferrous complex with ferene, and the corresponding colour analysis described herein.

We believe the work presented here are the first steps toward a powerful technology in iron metabolism tested. We have been inspired by the capillary blood glucose monitor (BGM) which currently serves to screen for glucose metabolism anomalies, as well to provide evidence for short-term intervention. These and all other successful capillary blood analysis tools require user-friendly capillary blood self-drawing technology to obtain 20 – 30 μL samples for approximately 1-minute processing and rapid analysis (reading in less than 2 minutes). Measuring iron ion, and eventually when paired to measured TIBC and ferritin, appears to tightly fit the BGM analogy, wherein iron deficiency or overload can be screened. Further, when accompanied by a professional, diseases can be prevented or treated under the convoluting conditions of dietary changes, menstruation and pre- or post- blood donations (therapeutic or philanthropic). Last, the integration with a smartphone allows rapid data processing, storage and transmission, which maximizes the benefit of applying mobile platforms to health applications.

Conflicts of interest

There are no conflicts to declare.

References

1. L. A. Kaplan, A. J. Pesce, *Clinical Chemistry Theory Analysis and Correlation*, Mosby Inc, 1996.
2. Nutritional anaemias: tools for effective prevention and control. Geneva: World Health Organization; 2017. Licence: CC BY-NC-SA 3.0 IGO.
3. E. Beutler, A. V. Hoffbran, J. D. Cook, *ASH Education* , 2003, **1**, 40-61. doi:10.1182/asheducation
4. R. J. Stoltzfus, *Food and Nutrition Bulletin*, 2003, **24**, S99-S103. DOI: 10.1177/15648265030244S206
5. M. Worwood, *Clin Chim Acta*, 1997, **259**, 3-23.
6. B.M.N de Graaff, A. Neil, K. Sanderson, Y. Kristy, K.C Yee, A. Palmer, *Journal of Australian Health Review*, 2017, **41**, 254-267.
7. Iron Disorders Institute, Hemochromatosis Diagnosis Algorithm, 2011. Website:http://www.irondisorders.org/Websites/idi/files/Content/854256/HHC%20ALL2011.pdf
8. A. Merryweather-Clarke, J. Pointon, J. Shearman, K. Robson, *J Med Genet*, 1997, **34**, 275–278.
9. R. Gozzelino, P. Arosio, *The Importance of Iron in Pathophysiologic Conditions*, Frontiers Media, Lausanne, 2015.
10. P. Adams and J Barton, *Blood*, 2010, **116**, 317-325
11. S. K. Vashist, *Biosensors*, 2017, **7(4)**, 62.
12. Statista, Number of smartphone user worldwide from 2014 to 2020, 2016.
13. B.Coleman, C. Coarsey, W. Asghar, *Analyst*, 2019, **144**, 1935.
14. S. Knowlton, A. Joshi, P. Syrrist, A. F. Coskun, S. Tasoglu, *Lab on a Chip*, 2017, **17**, 2839
15. Xu Xiayu, et al. Electrophoresis, 2019, 40(6), 914-921
16. Sensors and Actuators B: Chemical, 2018, 276, 48-56
17. Wiener laboratories, Direct colorimetric method for iron determination in serum or plasma.
18. Elitech Clinical Systems, Iron ferene, 2016. URL: https://vitaldiagnosticsinc.com/support/ifu/FTNA-FEFE-v3_29-03-2016.pdf
19. J.D. Artiss, S. Vinogradov, B. Zak, *Clinical Biochemistry*, 1982, **14(6)**, 311-315
20. T. Hirayama, H. Nagasawa, *Journal of Clinical Biochemistry and Nutrition*, 2017, **60**, 39-48.
21. Mohammad, A. Nejad, M. Wabler, H. Zhou, J. Mihalic, C. Gruethner, T. DeWeese, R. Ivkov, *International Journal of Hyperthermia*, 2018, **34 (4)**, 373-381.
22. Bio-Rad Laboratories, online website: http://www.bio-rad.com/en-us/product/blot-absorbent-filter-paper?ID=1172dcb6-635f-46e3-b0db-76ccbb39413e
23. P. Ponka, M. Tenenbien, J. W. Eaton, *Handbook on the Toxicology of Metals*, 2007, 30, 577-598.
24. D. J. Hennessy, G. R. Reid, F. E. Smith, S. L. Thompson, *Canadian Journal of Chemistry*, 1984, **62**, 721-724.

25. A. Shrivastava, V. B. Gupta, *Chronicles of Young Scientists,* 2015, **2,** Issue 1. DOI: 10.4103/2229-5186.79345
26. N. Saadati, M. P. Abdullah, Z. Zakaria, S. B. T. Sany, M. Rezayi, H. Hassonizadeh, *Chemistry Central Journal,* 2013, **7**, 63.
27. D. Zhao, X. Xian, M. Terrera, R. Krishnan, D. Miller, D. Bridgeman, K. Tao, L. Zhang, F. Tsow, E. S. Forzani, N. Tao, *Clinical Nutrition*, 2014, **33**, 341-347
28. E. Macias, A. Suarez, J. Lloret, Sensors, 2013, 13, 17292-17321, doi:10.3390/s131217292
29. G. Girardin, A. Bonnabel, E. Mounier, Technologies & Sensors for the Internet of Things, 2014 (online report)

571064 Colorimetric Detection of Methicillin-Resistant *Staphylococcus Aureus* Using DNA Nano-Sensing Probes

Colorimetric Detection of Methicillin-Resistant *Staphylococcus aureus* Using DNA Nano-sensing Probes

Y.M. Garcia[1, 2], Y.C. Tung[1], H. Sakamoto [3], and H.S. Chuang[1, 4, *]

[1]Department of Biomedical Engineering, National Cheng Kung University, Tainan, Taiwan

[2]School of Chemical, Biological, Materials Engineering and Sciences, Mapua University, Manila, Philippines

[3]Tenure-Track Program for Innovation Research, University of Fukui, Fukui, Japan

[4] Center for Micro/Nano Science and Technology, National Cheng Kung University, Tainan, Taiwan

ABSTRACT

According to the World Health Organization (WHO), 20% of all documented *S. aureus* infections are attributable to Methicillin-resistant *Staphylococcus aureus* (MRSA), but figures rise up to 80% in some developing countries.[1] MRSA is still considered as one of the major threats amongst hospital-acquired infections globally, causing lots of life-threatening diseases, such as complicated wound abscesses, pneumonia, endocarditis, and sepsis[2]. Especially for those immunocompromised patients who are vulnerable to bacterial infections, this may lead to higher mortality since approximately 5% of patients in the hospitals are MRSA-carrier according to the U.S. Center for Disease Control and Prevention's report[3]. As a result, preventing the spread of the disease and initiating a timely treatment with an appropriate antibiotic, i.e. a simple, fast and sensitive method of detection, becomes vital for saving more lives. In this study, we present a rapid diagnostic for MRSA infection by targeting the organism's biomarker, *mecA* gene. Detection of *mecA* gene in the sample is tantamount to the presence of the resistant strain of *Staphylococcus aureus* (SA) and that can selectively distinguish MRSA from methicillin-sensitive *Staphylococcus aureus* (MSSA) [4]. The nano-sensing probes comprise a mixture of two designed DNA probes, DNA Probe I and DNA Probe II, targeting a specific sequence in two different regions in the MRSA *mecA* gene. The DNA Probe I is conjugated with carrier polystyrene (PS) particles (d_p= 50 μm) while the DNA Probe II is conjugated with gold nanoparticles (AuNPs). Pretreated liquid samples containing the extracted MRSA *mecA* gene together with the designed DNA sensing probes are allowed to mix and hybridized in the chamber for a few minutes. After the reaction is done, the mixture will be pumped through the filter adjacent to it that serves as an obstacle for the aggregated hybridized PS/Probe I-MRSA *mecA* gene-Probe II/AuNP complex but will allow other small particles (AuNPs and DNA) to pass through. When the MRSA *mecA* genes are present, the two DNA Probes bind to its complementary DNA in the MRSA *mecA* gene, forming the PS/Probe I-MRSA *mecA* gene-Probe II/AuNP complex. MRSA *mecA* gene acts as a linker between the two DNA sensing probes. As a result, the PS/Probe I-MRSA *mecA* gene-Probe II/AuNP complex tends to accumulate in the filter paper, resulting in visible red color due to the aggregation of AuNPs. Conversely, in the absence of MRSA *mecA* gene, hybridization and linkage of the two DNA sensing probes are not possible, thus allowing the AuNPs to be filtered out and washed away from the filter paper. The absence of AuNPs then results in no visible red color in the filter paper. In summary, we proposed an integrated platform combining DNA nano-sensing probes that can detect a specific target and that its presence can easily be evaluated by naked eye, prompting its simple accessibility in point-of-care applications. This platform can address current challenges in rapid MRSA diagnosis with high sensitivity and specificity through the detection of MRSA *mecA* gene. Likewise, this platform can be extended in the diagnosis of other bacterial diseases such as Mycobacterium tuberculosis by applying the same principle of detecting a specific target using designed DNA nano-sensors.

Keywords - Methicillin-resistant *Staphylococcus aureus*, *mecA* gene, DNA nano-sensing probes, colorimetry.

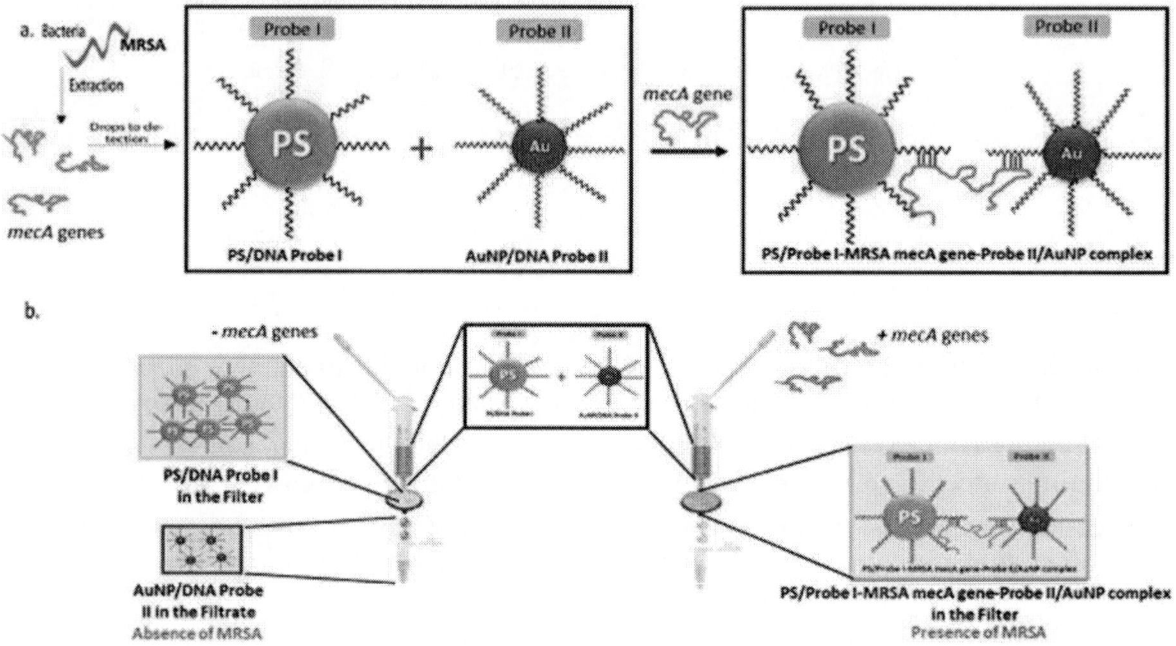

Figure 1: Scheme and experimental setup. (a) Schematics. The extracted *mecA* gene from the samples is allowed to react and hybridized with two kinds of DNA nano-sensing probes Probe I and Probe II. The *mecA* gene act as a linker between the two probes forming the PS/Probe I-MRSA *mecA* gene-Probe II/AuNP complex. (b) In the presence of the target *mecA* gene in the samples, there will be the hybridization of the DNA Probes I and II with *mecA* gene forming the PS/Probe I-MRSA *mecA* gene-Probe II/AuNP complex that aggregates and are trapped in the filter paper after filtration. Conversely, if the *mecA* gene is absent, then no complex will be formed, allowing the small AuNP/DNA Probe to pass through the filter and red filtrate is observable due to the color of AUNP that was filtered out.

References:

1. WHO. Antimicrobial resistance: global report on surveillance 2014. http://www.who.int/drugresistance/documents/surveillancereport/en/on

2. CDC. Healthcare-associated Infections: Staphylococcus aureus in healthcare settings http://www.cdc.gov/HAI/organisms/staph.html

3. CDC. Methicillin-resistant Staphylococcus aureus: General information. https://www.cdc.gov/mrsa/community/index.html

4. Watanabe K, Kuwata N, Sakamoto H, Amano Y, Satomura T, Suye S (2015) A smart DNA sensing system for detecting methicillin-resist *Staphylococcus aureus* using modified nanoparticle probes. Biosens Bioelectron 15:419–423

568792 Developing Janus Particles Enabled Rotational Diffusometry for High-Sensitive Diabetic Retinopathy Detection

Developing Janus Particles Enabled Rotational Diffusometry for High-Sensitive Diabetic Retinopathy Detection

Wei-Long Chen[1], and Han-Sheng Chuang[1,2] *

***(oswaldchuang@mail.ncku.edu.tw)**

[1] Department of Biomedical Engineering

National Cheng Kung University, Tainan 701, Taiwan.

[2] Center for Micro/Nano Science and Technology

National Cheng Kung University, Tainan 701, Taiwan.

Abstract- Brownian motion is a natural phenomenon in which tiny particles display random motion in fluids. Nevertheless, the diffusion of particles is proportional to temperature and viscosity but inversely proportional to radius of particles according to the Stokes-Einstein equation [1]. Based on this principle, diffusometry has been applied to detect fluid temperature, fluid viscosity, and particle dimeter. Recently, this technique has been extended to biomedical fields, such as investigation of microorganism motility [1] and rapid diagnosis of diseases with specific biomarkers [2]. Despite the above mentioned advantages, the conventional diffusometry is incapable of detecting low-abundance analytes. Unlike translational Brownian motion, rotational Brownian motion emerges a more sensitive response to the particle size. According to Stokes-Einstein Debye equation [3], the rotational diffusion of the particles is inversely proportional to the cubic radius of the particles under the consistent temperature and viscosity of fluid. In this paper, we aimed to diagnose DR with a selective biomarker, tumor necrosis factor alpha (TNF-α), in the early diagnose. Past literature [4] has reported that TNF-α is one of the biomarker proteins associated with DR in human tear. In this research, we developed a novel technique for sensing target proteins based on rotational Brownian motion using Janus particles. Only a small sample volume (~2 μL) was required (Figure 1A). Janus particles were fabricated by coating half of fluorescent polystyrene (PS) beads with a gold film, and purified with a disk filter. Taking the advantage of the characteristic of half gold/ fluorescence surface, the rotational Brownian motion signal was expressed through blinking intensity. In the preparation of functionalized particles, polyclonal antibodies were conjugated with functional submicron-sized PS particles by EDC and NHS, while monoclonal antibodies were conjugated on the gold side of Janus particles. As TNF-α was present in the sample, submicron-sized PS particles would attached to Janus particles to form a sandwich immunocomplex, resulting in a change of rotational Brownian motion. A parameter, correlation time, was used to quantify the cross-correlation intensity of Janus particles [5]. The correlation time was observed to significantly increase with TNF-α. The results showed that the correlation time of the immunocomplexed Janus particles was longer than that of pure Janus particles. The present limit of detection of the technique could achieve 10 ng/mL (Figure 1B). The Janus particles enabled rotational diffusometry was herein proven to be a prospective diagnostic for more other protein detections in the future.

Keywords: Janus particles, TNF-α, Rotational Brownian motion, Protein detection, Biosensor, Immuno-sensing, Diffusometry

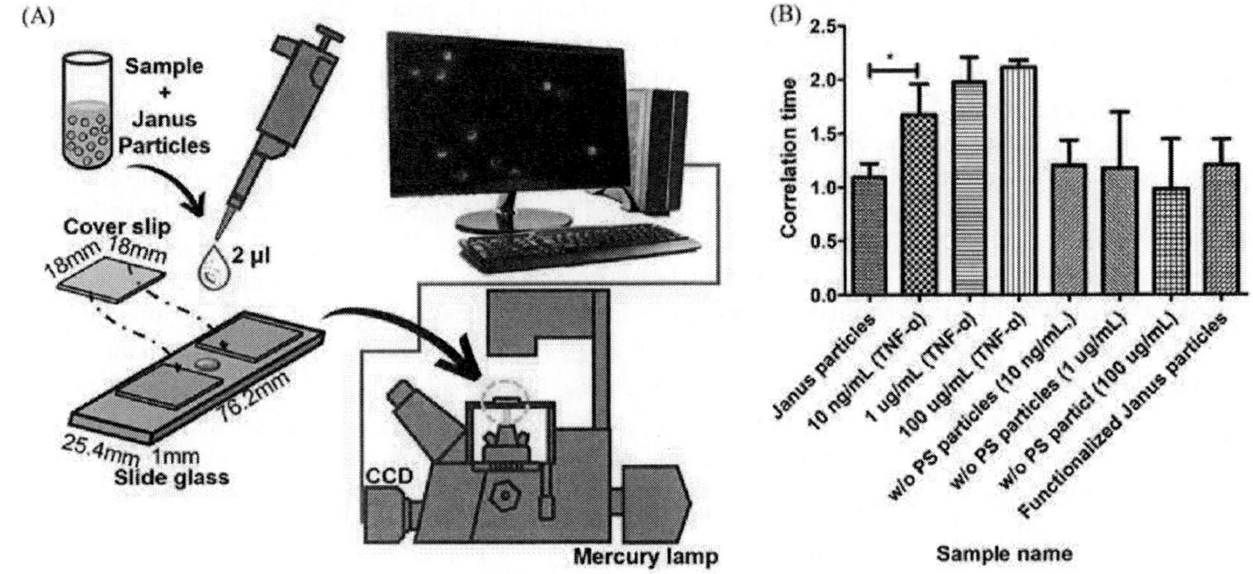

Figure 1: The experimental setup and results. (A) The experimental schematic. TNF-α antigen and functionalized submicron size PS particles were incubated with functionalized Janus particles to form the immunocomplex. Rotational Brownian motion could be examined under microscope. Correlation time could be estimated from the correlation intensity of Janus particles by computer. (B) The correlation time of different condition. The result show that the significantly increase with concentration of TNF-α. There is no significant difference when lack of submicron PS particles and TNF-α antigen compare to the pure Janus particles.

Reference:

1. Einstein, A., Investigations on the Theory of the Brownian Movement., *DoverPublications,* 1956.
2. Chung, C.Y., Wang, J.C., Chuang, H.S., Rapid Bead-based Antimicrobial Susceptibility Testing by Optical Diffusometry,, *PLOS One*, 2016.
3. Chen, Y.J., Cheng, H.P., Chuang, H.S., Enhanced Diffusometric Immunosensing with Grafted Gold Nanoparticles for Detection of Diabetic Retinopathy Biomarker Tumor Necrosis Factor-£\., *Biosensors and Bioelectronics*, 2018.
4. Lavalette, D., Tetreau, C., Tourbez, M., & Blouquit, Y., Microscopic viscosity and rotational diffusion of proteins in a macromolecular environment., *Biophysical journal*, 1999.
5. Costagliola, C., et al., TNF-Alpha Levels in Tears: A Novel Biomarker to Assess the Degree of Diabetic Retinopathy., *Mediators of Inflammation*, 2013.
6. Chen, C.J., Chen, W.L., Phong, P.H., Chuang, H.S., Investigation of Micro-volume Viscosity with Janus Microbeads Based on Rotational Brownian Motion., *Sensors*, 2019.

576942 Early Screening of Infection Using Electrochemical Point-of-Care Biosensor

Infectious disease is ranked one of the top ten leading causes of death worldwide according to World Health Organization (WHO). Furthermore, about 16.8 million hospital visits have been registered in 2018 owing to infectious diseases as per Center for disease Control & prevention (CDC) reports. The primary cause of infection is attributed to bacterial or viral source. Despite medical advancements in health care technology, there exists a major challenge to distinguish between bacterial and viral infections. It is essential to identify if the infection is bacterial or viral as it dictates the course of treatment accordingly. Clinical ambiguity propels for antibiotic misuse, with deleterious repercussions to the patient and health care system including the advent of antibiotic resistance. Current diagnostic tests assist in determining infection etiology, but endures with limitations including prolonged result time, large sample volumes, uncertain clinical interpretations and false alarms due to non-specific techniques. Owing to the pressing need for a viable solution, we have developed a rapid point-of-care biosensor to screen for infection utilizing low plasma sample volume of 40 μL. Enabling a unique signature with the combination of specific biomarkers such as tumor necrosis factor related apoptosis-inducing ligand (TRAIL), interferon γ-induced protein-10 (IP-10) and bacterial induced C-reactive protein (CRP) that facilitates appropriate treatment approach. The developed non-faradaic label-free biosensor using electrochemical impedance spectroscopy (EIS) technique quantifies the target biomarker concentration by measuring impedance response due to binding interaction of the target biomarker with the specific capture probe. The biosensor has a detection limit of 50 $pgmL^{-1}$ for IP-10, 1 $pgmL^{-1}$ for TRAIL and 0.2 μgmL^{-1} for CRP in plasma. The developed biosensor does not respond to any non-specific interactions. This novel biosensor allows sensitive detection of all three biomarkers in plasma with a response time of 5 minutes devoid of the need for sample dilution. Attributing to the specific detection methodology the developed point-of-care biosensor could serve as an imperious clinical resource to screen for infection.

566654 Sensing Mechanisms of Floating Gate Transistors

Transistors have been utilized extensively for biodetection due to their fast and sensitive signal transduction. The Electrolyte-Gated Transistor (EGT) is a variant of the field effect transistor where the conventional gate dielectric is replaced with an electrolyte. The EGT is especially attractive for biosensing applications because it offers the added advantage of low voltage operation, and can be fabricated using fast and potentially inexpensive manufacturing techniques such as printing. The Floating Gate Electrolyte-Gated Transistor (FG-EGT, or FGT) incorporates a floating gate that physically separates and electronically couples the active sensing area with the transistor. This design eliminates degradation and optimization issues associated with interactions of the semiconductor/dielectric with the analyte medium, and allows for an independent choice of electronic materials and capture chemistry. The floating gate therefore allows this device to overcome many drawbacks of traditional transistor-based sensors that directly utilize the semiconductor or dielectric surface for molecular capture. Previous work has shown the utility of FGTs for the fast and reliable detection of ssDNA in buffers, ricin in potable liquids, and gluten from different grain sources.

In this work, self-assembled monolayers of MUA and alkanethiols were utilized to systematically alter the charge state and capacitance of the sensing surface, demonstrating the ability of the device to detect changes in both interfacial charge and capacitance. Further, a simple inverter amplification circuit was used to obtain higher signals from the device when compared to previous work using transfer curves. By varying the sensing area along during these experiments, we were able to deduce an optimum area that maximizes the signal obtained from the device. This enables us to identify the best device architectures for the detection of charge-based and capacitance-based signals. The capacitance values obtained from our experiments are compared to literature and cyclic voltammetry for further confirmation of our findings. Our study provides general conclusions regarding testing protocols and sensing areas that are applicable to a range of transistor-based biosensors, which are a promising class of sensing devices for food safety and point-of-care diagnostics.

565195 Oriented Freeze-Casting Fabrication of Resilient Copper Nanowire-Based Aerogel As Robust Piezoresistive Sensor

One dimensional (1D) metallic nanowires-based flexible electronics have promising applications in strain and pressure detection. However, current research on piezoresistive materials are often hindered by robust mechanical performance, elastic-responsive conductivity and inefficient scale-up capacity. Here, we demonstrate a multi-scale fabrication strategy for three-dimensional (3D) copper nanowire-based aerogel with cost-effectiveness, directional and structural regulation and stable strain–electrical signal response via oriented freezing technique. Added polyvinyl alcohol binder dramatically enhances free-standing property and mechanical compliance of aerogel via gluing effect. Besides, this effect between nanowire and polyvinyl alcohol has been revealed by molecular dynamics simulations. Oriented aerogel displays far superior spring-back abilities than does non-oriented aerogel and exhibits good compressibility (undergoing a cyclic compression-recovery at 60%strain) and tunable piezo-resistance sensitivity (0.267 to 0.04 kPa^{-1}). Additionally, van der waals force contributes actively to the gluing effect on nanowire and polymer. The presented strategy opens up a new possibility for mass production of nanowire-based aerogel with ordered pore orientation and remarkable force-sensitive response.

572076 Synthesis of Core-Shell Palladium Nanowires@ZIF-8 for Hydrogen Sensors

Core-shell palladium nanowires@ZIF-8 nanostructures were prepared by dispersing the palladium nanowires in ethanol and growing a shell structure of ZIF-8 over them. The synthesized core-shell structure was vacuum dried and calcined in an inert atmosphere to remove PVP on the surface. The structure had a BET surface area exceeding 600 m^2/g and was thermally stable to at least 350 C. A hydrogen sensor was fabricated from the core-shell composite material and tested. The high aspect ratio palladium nanowire core improves response time and room temperature operation whereas the outer shell of ZIF-8 may enhance selectivity and poisoning resistance by sieving larger reactive gas molecules like oxygen, methane, and other interferents.

572825 Charge-Transfer Salt Nanowire Electrocrystallization for Ammonia Detection

Affordable sensors with better performance are required in different fields, such as environmental monitoring, safety, and homeland security. Nanowire sensors are considered as the promising next-generation sensors. Major technological gaps between the basic research and sensor market include the high cost and complicated process for making nanowire sensors. In this talk, we will present our ongoing research on the substrate-mediated charge-transfer salt nucleation and its application on nano/micro wire sensor manufacturing. Our research showed that small particles can direct nucleation of charge-transfer salt nanowires including tetrathiafulvalene (TTF) and tetracyanoquinodimethane (TCNQ). TTF-TCNQ is often recognized as the first organic metal. Using electrochemical crystallization, TTF or TCNQ charge-transfer salts with different counterions, can be directly deposited onto photolithographic substrates. By simply changing the crystallization conditions and counterions, nano/micro wire sensors of various detecting capabilities can be made and assembled into sensor arrays for gaseous chemical detection. One of our target gases is ammonia, which is considered the second most hazardous industrial gas. Our sensors can detect as low as 1 ppm ammonia in less than a minute. Our sensor arrays show enhanced selectivity towards mixed gases such as the selective detection against humidity after applying the deep learning algorithm. This work has been supported by the National Science Foundation, US Army Research Office, and MTRAC.

575134 Synthesis and Characterization of Charge Transfer Complex Micro/Nanowire Junctions and Their Application in Gas Sensing

Micro/nanowire junctions are of important technological interest due to their unique interfacial and electric properties. Junctions are the transport bottlenecks in network-based micro/nanowires, understanding the properties of these junctions and their connectivity is crucial to understanding and controlling electrical conduction through these networks. In this talk, we show the results of the electrochemical growth of conductive micro/nanowires of charge transfer complexes (CTC). CTC wires are grown on adjacent microelectrodes by applying periodically interchanged potentials until connections between the electrocrystallized wires are formed. The morphology and electric properties of the junctions are characterized. The findings are used to fabricate new devices for gas sensing.

559511 Biomimetic Properties of Modified Graphene Oxide and Graphitic Carbon Nitride Nanomaterials for Biosensing Applications

Biomimetics nanomaterials are gaining tremendous interest as biosensors due to the ease in their synthesis, easy modification and tuning of their structure and active sites, high stability and sensitivity. They are replacing enzymes as enzymes are high cost, less stable and difficult in handling. Further the research on the biomimetic nanomaterials is also focused to ease in handling and simplification of monitoring for point of care detection. Since the discovery of intrinsic peroxidase like activity of magnetite nanoparticles by Lizeng Gao et.al, in 2007, hunt for oxidase and peroxidase like activity of other nanomaterials started. In the quest of finding new materials, with superior biosensing properties, we started working on the study of biomimetic properties of metal oxides, carbon nitrides and graphene oxide. Graphitic carbon nitride is a hotspot in the material science due to its intrinsic chemical properties and have been used in catalysis, photocatalysis and in photoelectronics. It is also a promising material for modification to increase in electrical properties for making it more useful in its efficiency as a catalyst for different applications. Furthermore, since discovery graphene and its derivatives have been extensively used by researchers for different applications. In our research we used graphene oxide and carbon nitride as peroxidase mimic. We doped them different dopants to increase their electrical and surface properties by changing their particle size, surface area and chemical reactivity. Characterizations of the materials were done through various physical characterisation and techniques, to study the structure, morphology and band gap changes of samples. Peroxidase like activity was evaluated through hydrogen peroxide detection. Optimization of different condition and parameter were performed in order to get the best results. The results of our modified materials turned out to be very positive in terms of detection of hydrogen peroxide and other analysts.

571154 Fabrication of Laser Printed Microfluidic Paper-Based Analytical Devices for Point-of-Care Applications

Fabrication of laser printed microfluidic paper-based analytical devices for point-of-care applications

Rajesh Ghosh[1*], Saranya Gopalakrishnan[1], Rangasamy Savitha[1], Thiruvengadam Renganathan[1], Subramanium Pushpavanam[1*]

[1]Department of Chemical Engineering, Indian Institute of Technology, Madras, Chennai, 600036, India

*Contact: ghoshrajesh1226@gmail.com, spush@iitm.ac.in

Microfluidic paper-based devices (μPADs) have emerged as a versatile platform for low- cost, portable sensing in a broad spectrum of fields such as disease diagnosis, food safety, environmental monitoring and drug screening. However, fabrication of these paper-based devices with selective wetting behavior i.e. hydrophobic and hydrophilic patterns pose several challenges. Currently, a number of fabrication strategies exist, such as, photolithography, wax printing, ink-jet printing and etching, screen printing, plasma treatment, flexographic treatment, chemical vapor deposition, laser treatment, hand held or automated tool etc. Utilization of these methods involve complex fabrication steps and use of sophisticated or rare machinery which are unavailable in most conventional laboratory settings. Herein, we propose for the first time, a facile technique to fabricate μPADs using a ubiquitous office laser printer. We use the inherent hydrophobicity of the laser printer ink to create hydrophobic fluidic barriers, without modification of factory settings of the printer. In this technique, microfluidic channels or sensors are fabricated by just printing using the laser printer and heating the printed substrate for generation of hydrophobic barriers. The prepared devices are rendered selectively hydrophobic/hydrophilic with flow of reagents and samples occurring by passive capillary action. The proposed method circumvents the need for clean-room facility, fabrication masks, expensive or harsh chemicals. The fabricated devices are demonstrated for their application in sensor development by performing the detection of nitrite and food borne pathogens. The proposed method will provide a versatile platform for rapid prototyping of μPADs especially in resource constrained regions.

Keywords: paper-based microfluidics, laser printing, point-of-care diagnostics

576442 Modeling Fermi Levels in Metal Functionalized TiO2 Sensors for Applications in Volatile Organic Biomarker Detection Associated with Pneumonia

There is a large knowledge base of characteristic volatile organic biomarkers (VOBs) that are detectable as a direct result of a particular disease. VOBs can be detected from many sources including patient breath and urine. The fabrication and testing of VOB sensors for the detection of various diseases is the topic of several ongoing projects in our research group. These VOB sensors utilize a highly specific interaction between a known VOB, a metal-functionalized titanium dioxide sensor, and an applied voltage. When designing a novel sensor for a different disease, it can be extremely difficult to determine the appropriate components because the correct choice for VOB, metal, and voltage are all of great importance and each component comes with many possible options.

A simplified version of the sensor system has been examined with various combinations of VOBs, metals, and voltages using modeling software that employs density function theory (DFT). DFT calculations were performed to determine the fermi level of TiO2 functionalized with a group of metals of including Co, Ni, Fe, and Cu. The fermi level was then determined for the selected system in the presence of an applied voltage and a known VOB. For a potential pneumonia breath sensor these VOBs included heptane, acrolein, and ethylbenzene. The more negative the fermi level became when the metal-functionalized titanium dioxide was in the presence of a particular VOB, the better the sensor was predicted to respond. Promising systems were then modeled to determine for selectivity by calculating the fermi level of the sensor in the presence of compounds commonly found in breath that are not indicative of the disease. The results from this study directly apply to the fabrication of a potential pneumonia breath sensor, but are also appropriate for a broad range of applications where volatile organic compound detection is needed.

564587 Factors Affecting CO_2 Accumulation in the Motor Vehicle Cabin

Unmitigated carbon dioxide (CO_2) accumulation within vehicle cabins may pose risks to drivers due to cognitive impairment caused by high CO_2 concentrations, e.g. exceeding 1,000 ppm for several minutes of exposure. This work investigates the effects of air recirculation mode and vehicle speed on CO_2 buildup profiles within a mid-size sedan vehicle (most popular category of cars sold in the United States). The CO_2 concentration was measured continuously with a double beam-based infrared pre-calibrated monitor with 10 s resolution inside a vehicle driven at different speeds over periods of up to 50 min. The use of recirculation mode resulted in the accumulation of CO_2 to levels above 1000 ppm in less than 10 minutes. Vehicle speed was shown to have a moderating effect on CO_2 buildup, which was correlated with increased effective air exchange between the vehicle cabin and the surrounding environment. A CO_2 management strategy was evaluated to mitigate the vehicle occupants' exposure to high CO_2 levels resulting from the use of recirculation mode. This strategy aimed to maintain CO_2 concentrations at a safe level within the vehicle cabin by turning on and off the recirculation mode, which enabled the control of indoor air quality within the vehicle cabin. The driver's energy expenditure (in kcal/day) was estimated from the experimentally determined CO_2 generation rates. A box model was used to simulate CO_2 concentration profiles when multiple occupants were present in the vehicle, and to predict the effects of vehicle speed, effective air exchange rate and driver's energy expenditure in different scenarios, providing broader insights on the potential risks associated with CO_2 level accumulation within vehicles.

573723 Curating Metal-Organic Frameworks to Compose Robust Gas Sensor Arrays

Metal-organic frameworks (MOFs) - tunable, nano-porous materials - are alluring as recognition elements in gas sensors. Mimicking the human olfactory system, an array of cross-reactive, MOF-based sensors could allow analyte detection in complex and variable gas mixtures containing confounding/interfering gas species. Herein, we address the question: given a set of MOF candidates and their adsorption properties, how do we select the optimal subset to compose a sensor array that accurately and robustly predicts the gas phase composition under measurement noise?

Using experimental adsorption data, we choose the optimal subset of MOFs to coat quartz crystal microbalances (QCMs), to compose a sensor array to determine the concentration of various analytes in the gas phase. The sensor array can be viewed as a function which maps a change in *gas composition space* to *sensor array response space*.
We generalize how a change in composition space affects the response generated by the sensors in response space. Different subsets of MOFs to form a sensor array can be compared by quantifying the magnitude of the response so as to distinguish among gas mixtures and diminish the influence of experimental measurement noise.

557182 Probe Design for NMR of Chemical Reactions at High Temperature and Pressure

We report here nuclear magnetic resonance (NMR) apparatus for the study of chemical reactions at elevated temperature and pressure in situ and in real time. NMR is a measurement that is frequently used to evaluate the chemical structure of a sample. In addition to chemical structure, other parameters are also available such as diffusion and molecular relaxation. An interesting application for NMR is to follow a chemical reaction from changes in the NMR spectrum. To do this, elevated temperatures and pressures are often required. This presentation will discuss two probe designs that allow us to follow reactions at elevated conditions, opening a window to refinery and commodity chemical processing reactions.

The first probe is a titanium alloy vessel with an internal free-piston. The liquid sample is loaded into a ceramic sample tube with the free-piston sealed to the end of the ceramic. The free piston equalizes the pressure inside and outside of the ceramic so the pressure is held only by the titanium vessel and not the ceramic tube. The titanium holds the argon pressurizing gas and is non-magnetic. The radio frequency (rf) coil is inside the titanium, wrapped directly around the sample holder for maximum signal. There are also two heaters inside the titanium vessel, to heat the ceramic vessel and sample. This probe has been tested to a temperature of 600 °C and a pressure of 40 MPa.

An advantage of the titanium probe is independent control of temperature and pressure. The probe is pressurized with argon from a high pressure argon cylinder. The heaters are controlled by a temperature regulator regulating at a thermocouple sealed inside the titanium vessel. With this design, the sample must be loaded as a liquid at ambient conditions.

In the second probe, the sample is held in a sealed glass capillary. The current capillaries are 7 mm OD, 2.6 mm ID. The sample is heated with forced air flowing over a wire-wound heater situated below the sample. The temperature is controlled by a regulator sensing a thermocouple located at the top of the heater. This probe can run up to 400 °C and the capillaries are routinely able to hold up to 7 MPa. The pressure is determined by the temperature, as the sealed samples are at constant density. An advantage of the glass capillary is simplicity. All of the pressure-holding task is the "responsibility" of the sealed capillary. Additionally, the sealed capillary approach provides the option to prepare samples that are both liquid and gas at atmospheric conditions. For example, we can load liquid oxygen or liquid carbon monoxide using liquid nitrogen to cool the tube while it is flame-sealed. Hydrogen gas can be provided from metal-hydrides that release hydrogen at reaction temperatures. Thus, liquids as well as normally-gaseous samples (O_2, CO or H_2) can be loaded.

Considering a two phase sample, liquid and gas, at room temperature, NMR experiments can be acquired on either the liquid or on the vapor by moving the sample tube within the rf coil of the probe.

NMR measurements of the chemical spectra at increasing temperatures and/or pressures can show formation of new chemical species. Measurements of the diffusion coefficient can give insight into the phase of the fluid or solution, with a much faster diffusion coefficient for gas than liquid. Using these measurements, the phase transition at different values of temperature and pressure can be plotted. The T_1 relaxation of a sample can also be measured in this vessel and will, again, indicate the phase of the sample and can also give insight into molecular dynamics. All of these measurements can be acquired in liquid, gas or above the critical point, in the supercritical region.

These probes yield information about a reaction at process conditions, both high temperature and pressure. It is difficult or impossible to acquire these data with external analysis when sampling from the reaction vessel where cooling and depressurization of the sample are required before analysis of the products. Phase fractionation upon return to ambient conditions can result in solids precipitating and gases escaping from the external analysis. In addition, the in situ NMR measurements may capture transient chemical intermediates. In this presentation, the design of both probes and the types of data that can be acquired will be discussed.

573651 Sensing and Detection of Arsenic By Localized Surface Plasmon Resonance of Gold-Vitamin B12 Complex and a Device Thereof

Background: Arsenic contamination of water sources is a severe problem in various parts of the world. Drinking of arsenic containing water has serious consequences in terms of human health hazard and even death. In fact, because of its severe toxicity, the safe limit of arsenic in drinking water is only 10 ppb (parts per billion). In contrast, many contaminated sites have arsenic concentration upwards of that of hundreds of ppb. Thus, detection and measurement of arsenic concentration becomes a critical first step, in order to remove arsenic from water; so as to obtain safe, drinking water, which is fit for human consumption. In addition, because of the colourless, odourless and tasteless nature of arsenic present in water, it is a formidable challenge to have a handy and portable arsenic detection system, capable of measuring arsenic concentration fairly accurately and within minutes, down to the level of 10 ppb, or if possible, even for lower concentrations.

Principle: Our current research has resulted in the development of a sensor chemistry (between the gold-vitamin B12 complex developed by us and the arsenic contaminant), right up to the stage of a small device. In addition, we have achieved a high sensitivity and selectivity to arsenic concentration measurement, which can be detected from 100 ppb down to 10 ppb, quite accurately. This can be extended down to 0.1 ppb, based on the type of measurement device employed, using the same sensor chemistry and optical spectra based measurement-principle developed by us.

The detection and measurement is based on the reaction between - (i) a water-dispersible methylcobalamin (vitamin B12) containing gold nanoparticle solution, with (ii) a sample of arsenic containing water. Simple mixing of this in a cuvette results in a controlled aggregation of tiny gold nanoparticles of about 6-7 nm diameter, as arsenic binds to gold via our methylcobalamin sensor molecule. This translates into a shift in the local surface plasmon resonance of gold nanoparticle and gives an increase in its absorbance peak intensity, in the visible light range. This absorbance intensity is found to be very specific and selective to only arsenic and is proportional to the arsenic concentration in the water sample, from about 0.1 ppb upwards, up to 100 ppb.

Salient features of sensor: This effect and the mechanism behind it has been utilized by us to first have an on-board spectrometer system, and then further improve it to a small device (comprising of a LED light source and a photodiode), to help measure arsenic concentration very fast within minutes. This is in contrast to any unwanted delay (of 20-30 minutes) or use of other toxic chemicals, like mercury etc., which impedes other current testing kits.

Application: The sensor system can be used as a portable device to identify arsenic contamination and its value for different water-sources and water-bodies, very quickly. This is based on validation of results from our sensor-system with that of inductively coupled plasma mass spectrometric (ICP-MS) measurement of other synthetic and real-life environmental water samples.

572846 Invited: Sensor Applications in Bioengineering: Detection of Small Molecules within Complex and Three-Dimensional Biological Milieu

Bioengineering applications in regenerative medicine and single-cell analysis present challenging microenvironments for diagnostic and real-time detection. We have developed and applied sensors to detect small molecules in neuronal, cardiovascular, blood and biofilm systems. In particular, we have aimed to establish easy-to-use sensor platforms that do not interfere with biological processes. These sensors are reagentless and can be deployed in the presence of complex macromolecular solutions that readily foul, degrade or block many traditional sensing approaches. Herein, we describe our efforts towards detection of neuroactive molecules in the brain, hypoxia gradients in three-dimensional cellular scaffolds, and subcellular oxygen concentrations during electrophysiological and chemo-mechanical single cell experiments.

575136 Tissue-like Neurochemical Sensors Operating in Brain and Gut

Normal neurotransmission relies on a precise and delicate balance of chemical signaling mediated by the release of variety neurotransmitters. Among them, the monoamines including dopamine and serotonin are uniquely involved in the regulation of cognitive processes such as emotion, arousal and memory. Therefore, it is of great interest in both basic research and clinical studies to monitor the fluctuation of monoamine in both brain and peripheral nervous systems. As a very soft and complex biomaterial, the human brain is under constant motion and deformation because of different physiological dynamics, while GI tract is a series of soft organs with inherent motility, joined in a long and twisting tube from the mouth to the anus. As a result, rigid neurochemical sensors can cause inflammation to the brain tissue, and neurochemical sensing in awake animals have historically been impossible to traditional rigid and fragile devices.

Softness and stretchability in new-generation neuroengineering tools are essential to minimize the perturbation of motions inherent in biological systems and maximize the physiological fidelity. Here we develop a new neuroengineering platform, based on soft materials, that could allow access to the brain and entire enteric nervous system for neurochemical sensing. The new neurochemical sensor is based on graphene-elastomer composite, which offers chronic, sensitive, and selective sensing of dopamine and serotonin in the brain. Ex vivo studies shows that the soft sensor could accommodate the peristalsis movement of the intestine and continuously monitor the serotonin dynamics via thermal treatment. In vivo experiments also show that the sensor could be implanted in the colon of mice to examine the drug-stimulated serotonin response. The new neuroengineering platform described here can provide unprecedented insights on the microbiota–gut–brain communication in the context of mental disorders and therapeutic interventions, and can lead to new therapies for psychiatric disorders through GI tract without invasive surgery to the brain.

570825 Cell-Based Electrochemical Sensors for Environmental Pollutant Detection

Endocrine disrupting compounds are ubiquitous; they are found in many products intended for common use, including pesticides, plastics, and even some medications. Yet, these chemicals are implicated in health problems ranging from diabetes to cancer. We have developed a cell-based electrochemical strategy to rapidly and sensitively quantify these compounds. Our strategy measures the biological activity of chemically dissimilar xenoestrogens by harnessing the specificity of a human estrogen receptor. This receptor is expressed on the surface of *E. coli* as a component of an electrochemical sandwich assay. Our sensors require no specialized skills to implement and have enabled very sensitive detection (sub-ppb levels) of environmental pollutants. We have successfully detected compounds in from complex solutions, such as baby formula, and have measured estrogenic activity from unknown compounds leeched from a BPA-free baby bottle.

577305 Biomimetic Composites As Microfabricated, Flexible, Degradable Electrochemical Sensors

Biomimetic composites as microfabricated, flexible, degradable electrochemical sensors

- M. Xu, S. Pradhan, R.K. Pal, V.K. Yadavalli

Department of Chemical and Life Science Engineering, Virginia Commonwealth University, Richmond, VA

Biomimetic composites of naturally derived and synthetic polymers provide exciting opportunities to develop mechanically flexible, physiologically compliant, and degradable bioelectronics systems. In prior work, we have shown that naturally derived silk proteins and conducting polymers (specifically poly(3,4-ethylenedioxythiophene) :poly(styrene sulfonate) (PEDOT:PSS)) can form functional bioinks for sensing and other electrochemical applications. These inks combine the mechanical strength, biocompatibility and degradability of silk proteins, with the conductivity and chemical functionalization of the PEDOT:PSS. The fabrication and characterization of photopatternable, water-based conductive inks comprising PEDOT:PSS and synthesized photoreactive silk proteins was earlier shown [1] The presence of photoreactive groups permits a fully aqueous photolithographic strategy to form conductive microelectrodes on both rigid substrates as well as flexible silk films [2].

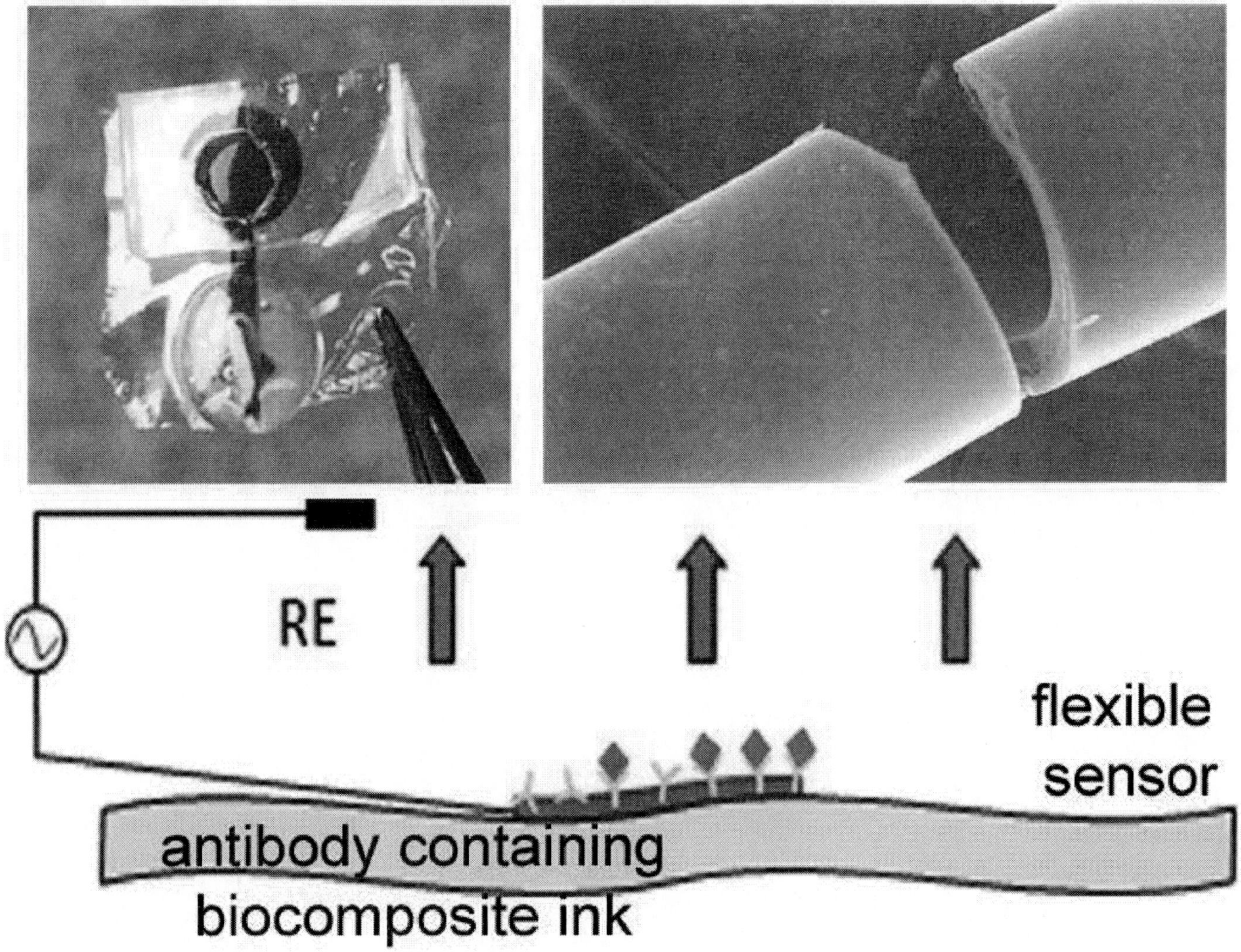

Here we will present the latest developments using this technology, showing how these devices can function for flexible bioelectronics applications in the form of microelectrodes for ultrasensitive detection, as well as components such as microsupercapacitors.[3, 4] We show how electroactive biomolecules such as neurotransmitters can be detected sensitively, while non-electroactive biomolecules such as glucose and glutamic acid, can be detected by encapsulating specific enzymes. Due to the room temperature, aqueous fabrication process, we can immobilize antibodies in these sensors resulting in flexible immunosensors for targets such as vascular endothelial growth factor (VEGF) for the first time [3]. The electroactivity of conductive ink can be improved by the addition of small amounts dopants. Using these doped composites, we further demonstrate flexible energy storage devices due to the capacitive nature of the biomaterial. The presence of silk proteins as the matrix of the composite makes it completely biodegradable and biocompatible, potentially resulting in transient or implantable devices. This platform can also be used for the specific control of cell culture in flexible sheets. The mechanical, biochemical and electrochemical characterization of the composite and its microfabrication are discussed. By virtue of a range of versatile properties, utility as bio-sensors, opto-electronic devices and flexible energy storage systems are envisioned.

References:

[1] "Photolithographic micropatterning of conducting polymers on flexible silk matrices" - RK Pal, AA Farghaly, MM Collinson, SC Kundu, VK Yadavalli, *Advanced Materials*, 28(7), 1406-1412, 2016

[2] "Conducting polymer-silk biocomposites for flexible and biodegradable electrochemical sensors" - RK Pal, AA Farghaly, MM Collinson, SC Kundu, VK Yadavalli, *Biosensors & Bioelectronics*, 81, 294-302, 2016

[3] "Flexible biosensors for the impedimetric detection of protein targets using silk-conductive polymer biocomposites" – M Xu, VK Yadavalli, *ACS Sensors*, in press, 2019

[4] "Fabrication of Flexible, Fully Organic, Degradable Energy Storage Devices Using Silk Proteins" - RK Pal, SC Kundu, VK Yadavalli, *ACS Applied Materials & Interfaces* 10 (11), 9620-9628, 2018

569635 Nanopore Based Highly Sensitive and Accurate Detection of HIV

Nanopore Based High Performance Point-of-Care Biosensor for Acute HIV Detection

Xiaojun Wei, Chang Liu

Human immunodeficiency virus (HIV) is the causative agent of acquired immuno-deficiency syndrome (AIDS). There are over 1.8 million infections HIV occur annually worldwide, and 30%¨C50% are attributed to recently infected individuals that are unaware of their disease state. Rapid diagnosis of acute HIV (within 3 weeks of infection) may reduce new cases by 10%¨C25%. The fourth-generation rapid diagnostic test for HIV is currently employed worldwide as the frontline screening test. This Enzyme-linked Immunosorbent Assay (ELISA) based test is designed to detect HIV-1 p24 antigen in addition to HIV antibodies to achieve early detection. However, several recent field tests have indicated unacceptably low diagnostic sensitivity (0%) of acute HIV-1.

Herein, a nanopore biosensing platform with single-molecule level sensing and analysis capacity has been developed to address the unmet need for rapid and accurate diagnosis of acute HIV to reduce transmission. In our approach, p24 antigens isolated from plasma samples were cleaved into several specific peptide sequences, which can be recognized and quantified by their signature signals during translocation through the nanopore. A portable QuantiPep-HIV prototype suitable for point-of-care settings was developed and tested with serum samples from HIV-infected and uninfected individuals to demonstrate the feasibility and to evaluate the diagnostic performance. Our results showed that QuantiPep-HIV can improve the diagnostic sensitivity and specificity of acute HIV.

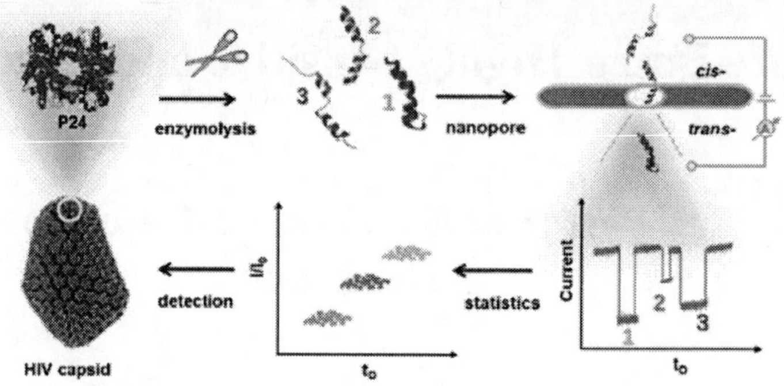

Figure. Schematic diagram for detecting peptide fragments from p24 antigens by nanopore

582378 Wireless, Soft, Skin-integrated Microfluidic Systems for Electrochemical/Colorimetric Biosensing and Capture of Sweat

Human eccrine sweat is a rich composition of nearly 60 known biochemicals including metabolites, electrolytes, vitamins, amino acids, exogeneous drugs, and small proteins. Advances in sensing technologies, conformal electronics, and materials science lay the foundation for recent demonstrations of wearable sweat sensors. Unfortunately, majority of the devices rely either on active, battery-powered electronics for electrochemical detection or passive, colorimetric chemistries for visual readout. Complex construction, large size/weight and possibility of battery leakage represent disadvantages of the former; semi-quantitative operation and limited scope of measurable biomarkers are drawbacks of the latter. Additionally, present wearable sensors can detect only a handful of biomarkers present in the sweat. Quantification of the remaining chemicals still requires sweat collection using adsorbent pads followed by off-site chemical analysis. Such a route provides average concentration of sweat analytes and restricts the acquisition of their temporal concentrations. This prohibits the understanding of the relation between dynamic variations in sweat chemistry and human physiology. In this talk, I will describe an unusual collection of enabling materials, sensing technologies, advanced microfluidics, stimulus dependent biocompatible power systems, galvanic time-tracking schemes, and wireless electronics to address these issues. Specifically, I will describe my strategies that include soft microfluidics, fuel cells, trigger-activated biocompatible batteries, colorimetrics and near field communication technologies that result in new classes of wireless, skin-integrated devices for detailed analysis of human sweat.

574301 The Electrochemical Impedance Response of a Continuous Glucose Monitor

The Gerischer impedance has been observed in systems with an electrochemical reaction coupled with a homogeneous chemical reaction. Under the assumptions of linear homogeneous reaction, equal diffusion coefficients and a semi-infinite condition, the Gerischer impedance can be solved analytically [1, 2]. However, with non-ideal electrochemical systems, where coupled nonlinear homogeneous reactions and various mass-transfer properties make exact analytical resolution impossible, a more detailed analysis is required.

In the present work, a mathematical model for the impedance response of glucose-oxidase based electrochemical biosensors has been developed [3]. The complicated coupling between thirteen homogeneous reactions and three heterogeneous reactions has been considered in the model. The homogeneous reactions included anomerization between α-D-glucose and β-D-glucose and four reversible enzymatic catalytic reactions transforming β-D-glucose and oxygen into gluconic acid and hydrogen peroxide, pH-dependent enzymatic activity and a biological buffer system. The electroactive hydrogen peroxide was considered to be reversibly oxidized or reduced at the electrode.

The mathematical model is solved numerically by using the finite-difference method and Newman's BAND algorithm [4]. The model demonstrates how the coupled non-linear homogeneous reactions effect the diffusion impedance, which has broadened the scope of the Gerischer impedance. The model can be used to explore the influence of various system parameters on limiting current, reaction profiles, and diffusion impedance. The system parameters, including interstitial glucose concentration, oxygen concentration, active enzyme concentration, diffusion coefficients, reaction rate constants and layer thickness, are related to various sensor working conditions such as body sugar level, inflammation, sensor degradation and sensor design.

References

1. H. Gerischer, "Wechselstrompolarisation Von Elektroden Mit Einem Potentialbes- timmenden Schritt Beim Gleichgewichtspotential," *Zeitschrift fur Physikalische Chemie*, **198** (1951) 286–313.
2. M. E. Orazem and B. Tribollet, *Electrochemical Impedance Spectroscopy* (John Wiley & Sons, Hoboken, NJ, 2017), 2nd edition, p. 279-293.
3. M. Gao, M. S. Hazelbaker, R. Kong, and M. E. Orazem, "Mathematical Model for the Electrochemical Impedance Response of a Continuous Glucose Monitor," *Electrochimica Acta*, **275**(2018),119-132
4. J. S. Newman and K. E. Thomas-Alyea, *Electrochemical Systems* (John Wiley & Sons, Hoboken, NJ, 2004), 3rd edition.

Acknowledgement

The support of Medtronic Diabetes (Northridge, CA) and Andrea Varsavsky, program monitor, is gratefully acknowledged.

579773 Leading the Industry 4.0 Journey in the Manufacturing Sector with Cross-Industry Collaboration and AI-Powered Innovation

While manufacturing represents 16% of global GDP, manufacturing industries account for 64% of global R&D spend. Yet innovation is long overdue since the manufacturing sector has experienced a decade of productivity stagnation and demand fragmentation. Organizations that have taken Industry 4.0 innovation to scale beyond the pilot phase have experienced unprecedented increases in efficiency with minimal displacement of workers. However, most companies appear to be stuck in "pilot purgatory". Widespread adoption of artificial intelligence (AI) and allied Industry 4.0 technologies at scale, through the combined efforts of companies and governments, can lead to a sizeable increase in global wealth creation, benefiting people throughout society.

However, there is potential for worker displacement if changes are not handled properly. Strengthened collaboration among stakeholders is needed to understand unexpected consequences and manage the transition.

Leaders in digital manufacturing and globalization 4.0 exhibit all the essential characteristics of the Industry 4.0 revolution. Moreover, they confirm the hypothesis that they have the potential to generate new economic value, driven by improvements in the full spectrum of production value drivers: resource productivity and efficiency, agility and responsiveness, speed to market and customization to meet customer needs. Transforming existing production systems, innovating value chains and building new business models with disruptive potential for established businesses can create value.

579424 Real-Time Analytics for IIoT

Digital transformation, connected enterprise, and IIOT are various names that refer to an increasing interest by the manufacturing industry to make better use of the data they generate. A key driver behind this heightened interest is the expectation that the business will benefit from mining actionable information out of that data.

The prevailing discourse today equates successful digital transformation with big data analytics (often delivered via a cloud-centric infrastructure). This talk will argue that, at least in industrial applications, there are a significant subset of problems for which a cloud-centered big data approach is not necessary.

This talk will present a family of embedded analytics capabilities that are designed to extract actionable information out of streaming operation data. A brief description of each capability along with the results from real-world applications of the technology will be presented.

579314 Process Optimization in the Age of Industry 4.0

Industry 4.0 represents a new trend of automation and data exchange in manufacturing technologies and fosters the concept of a "smart factory". Within this framework cyber-physical systems monitor physical processes, create a virtual copy of the physical world and make decentralized decisions. This talk provides an overview of modern optimization optimization tools and solution strategies within this distributed decision-making process. In contrast to current engineering tools which are based on simulation models and expand to optimization, we describe the development of more advanced decision-making tools that start with an optimization model and flesh out the details of the process. In particular, we highlight the development of equation oriented optimization platforms with fast solvers, advanced modeling environments and efficient interfaces between them. Moreover, we extend these features to NLP sensitivity, nonsmooth models and data-driven models. Finally, we explore the application of these tools for distributed optimization, optimization under uncertainty and the integration of design, operations and control.

580904 Value-Driven Digitalization Initiatives at Dow

Largely, the process industry has matured beyond the initial frenzy around automated reports, dashboards, and the embedding of statistics in software products, and has firmly moved toward building out infrastructure and decision-support systems to improve safety and maximize value.

Industry megatrends around M&A for narrowed domain focus, commoditization, low capital intensity, geo-political uncertainty, and emphasis on sustainability and customer experience require nimble and responsive supply chains, systems analysis for understanding trade-offs, and maximization of value from asset footprints. Further, with the emergence of AI platforms, the time has never been more ripe for practitioners to help an allegedly sluggish and slow-to-adopt industry to embed digital tools as a core part of their strategy.

A lot of emphasis has rightly been on external visibility of product portfolios to customers, as well as on internal decisions around what, when, and where to produce product---consequently involving IT, customer service/marketing, and supply chain functions. Although there continues to be significant opportunity in this space, AI technology and analytics has become increasingly relevant to R&D groups in the development of new products and innovative processes; and to maintenance and operations organizations in predicting and responding to process events.

Further, most digitalization projects fall in one of three areas of increasing implementation complexity: (1) technology that enables improved decision-making in a local environment that can be then cloned in different environments with small tweaks; (2) technology that is specific to a particular domain, business, class of products, or process; or (3) technology that spans across multiple assets and allows consolidated decision-making at scale. These three categories differ in the effort needed to standardize, scale, and get buy-in from stakeholders. These pose slightly different challenges, and the way organizations structure support organizations, empower employees, and manage programs is critical to the successful deployment of these systems and in effecting culture change.

Here, we provide specific examples where Manufacturing 4.0-related projects have enabled product, process, maintenance, and supply chain innovation at Dow, including recent success stories of deploying process intensification at scale. We also cover what Dow has done to intensify its organizations through three key capabilities---the Digital Marketplace, the Digital Operations Center, and the Digital Fulfillment Center, where domain experts and engineers sit alongside software architects, UX designers, and database experts to deliver solutions effectively. Further, we discuss how communities of practice around analytics have helped us align analytics professionals across different functions to develop a coherent strategy for the company around, for example, deployment best practices, building data pipelines, and digital twins. External partnerships with universities, consultants, and vendors do play a key role in technology transfer and adoption, and balancing the development of in-house talent with the right external support is important as well.

Through these stories, we hope to illustrate how an organization can plan a steady and firm climb out of the trough of disillusionment with digital technologies into the plateau of productivity.

581347 Chemical Engineering Innovations for a Renewable Economy

Chemical engineering evolved in the twentieth century with rise in the use of fossil resources and myriad of products derived from then. An array of these products required great innovations and made huge impact on the human race, and such examples include urea, transportation fuels, polymers, water purification chemicals, etc. Going forward as we make transition from fossil resources to renewable resources such as solar energy, wind and sustainably available biomass, we again have great opportunities for innovation that would not only make great impact but redefine chemical engineering. My talk will focus on my own experiences with challenges and opportunities presented to us as chemical engineers. I will share an overall perspective on the use of biomass as sustainable source of carbon, harvesting solar energy in conjunction with growing food without any land constraint, and creating solution processed solar cells that utilize chemical engineering principles along with knowledge from other engineering disciplines.

581350 Engineering Materials for Clinical Application

Dr. Schmidt's research has resulted in numerous patents that are licensed to companies, including AxoGen, Alafair Biosciences, Smith and Nephew, and Siluria Technologies. Comments for this panel will focus on peripheral nerve applications and one key success that is the foundation for the Avance Nerve Graft from AxoGen. In addition, along the way to uncovering strategies for nerve regeneration, Dr. Schmidt and her team also discovered a biomaterials processing approach that led to a separate application of protecting tendon's after surgery, resulting in the VersaWrap Tendon Protector from Alafair Biosciences, a start-up company in Austin, Texas.

Nerve Repair: Dr. Schmidt's technology for processing decellularized nerve tissue was licensed to AxoGen in Alachua, Florida (http://www.axogeninc.com) for the AVANCE^{TM} nerve graft. Many tens of thousands (~100,000) of AxoGen's AVANCE^{TM} grafts have been implanted in patients at hundreds of hospitals globally.

Post-surgical Wound Management: Using a novel in situ crystallization processing technique, Dr. Schmidt's team created mechanically robust hyaluronic acid-based hydrogel films that can be easily handled, sutured, and delivered laparoscopically. This research is the basis of Alafair Biosciences, a start-up company in Austin, TX (http://www.alafairbiosciences.com), which is run by Dr. Schmidt's PhD graduate and a neurosurgeon collaborator. Alafair's first product, VersaWrap^{TM} Tendon Protector, has been implanted in more than 500 patients since December 2017.

576965 Sequence-Specific Nucleic Acid Detection at <1 Am

A nucleic acid amplification-free, optics-free platform has been created for the detection of bacterial 16S rRNA at <1 aM (10^{-18}M). This sequence-specific nucleic acid (NA) detector is based on electromechanical signal transduction using a nanopore-based scheme. Our microfluidics device utilizes peptide nucleic acid (PNA) capture probes conjugated to submicron diameter polystyrene beads. Since PNA is charge neutral, the bead-PNA conjugates may be designed to be "charge neutral" until they hybridize target NA, at which point the complex becomes negatively charged and mobile in an electric field. If the electric field is oriented through a nanopore in a glass membrane that is too small for the bead-PNA conjugate with hybridized target to pass through, it will block the pore thereby causing an easily measured step reduction in ionic current. In this way, the selective, NA hybridization event is electromechanically transduced. An opposing electroosmotic flow through the glass pore provides an active mechanism to reduce false positives. No false positive signals were observed with *P. putida*RNA when this device was configured to detect *E. coli*16S rRNA. In fact, this device has proven capable of detecting *E. coli*16S rRNA at ~0.1 aM against a 1 pM background of RNA from *Pseudomonas putida*. Finally, the device readily detected *E. coli*at 10 CFU/mL in a 1 mL sample, also against a million-fold background of viable *P. putida*. These results suggest that this new technology may serve as the basis for small, portable, low power, and low-cost systems for rapid detection of specific bacterial species in clinical samples, food, and water.

576518 Detection of Water-Borne Pathogens Using a Smartphone-Based Portable Isothermal Amplification Platform

Waterborne pathogens (bacteria, virus, and protozoa) are significant contributors to the global health burden, resulting in an estimated 2.2 million deaths annually. Repeated disease outbreaks (*Escherichia Coli*, *Campylobacter jejuni*, *Salmonella*, *Vibrio cholerae*, *Giandia*, etc.) emphasize the dire need for a rapid and inexpensive diagnostic tools for reliable and efficient countermeasures. Although water treatment must continue to be at the forefront of such a countermeasure, it is critical to develop portable and inexpensive systems that scale for widespread deployment and implementation to directly detect the pathogens at the water sources (especially in resource-limited areas). We report an inexpensive and portable smartphone-based fluorescence imaging system for rapid detection of water-borne pathogens via Loop-mediated isothermal amplification (LAMP) of specific nucleic acid targets in disposable microfluidic chips. The microfluidic chips are machined with a desktop CNC machine with several optically clear pockets to house water samples. A low powered thermal management system ensures uniform heating of the samples for 30 minutes after which the fluorescence signals of the amplified targets are quantified with a high degree of confidence using a chromaticity-luminance analysis[1] to determine the presence/absence of pathogenic genetic material. Such low cost integrated nucleic acid amplification and fluorescence imaging system has the potential to simultaneously detect multiple pathogenic microorganisms. When implemented on a larger scale, such a technology could provide quantitative microbial risk assessment that evaluates the location-based likelihood of pathogen contamination resulting in rapid and reliable decision-making around such scenarios.

(1) Priye, A.; Ball, C. S.; Meagher, R. J. *Analytical chemistry* **2018**, *90*, 12385-12389.

568030 Cell Tracking Velocimetry: A Femtogram Resolution Fluorescence Cytometric Magnetometer

A magnetometer is an instrument that measures magnetism for various purposes, such as from compass in search of direction to the satellite magnetometers for surveying outer-space planets. Due to its simple mechanism, magnetism has found its use in the medical/life sciences field as well, such as magnetic resonance imaging technique, Miltenyl magnetic separation column, etc. Recently, instead of using magnetic particle conjugation like a lot of the traditional magnetic separation technique, various researchers have begun focusing on the intrinsic magnetization of biological entities. For example, Sun et al (2018) has tested the feasibility of magnetically separating unlabeled oxidized human red blood cells using microfluidic magnetic separation system. For another example, Park et al (2019) has found that glioblastoma cancer stem cells (CSCs) and non-stem tumor cells (NSTCs) have different magnetic properties due to their different iron-storage mechanism. Therefore, an ability to accurately measure/quantify the magnetic properties of biological entities has become very important.

Over the years, we have been developing an instrument referred to as the cell tracking velocimetry (CTV) which is a device that uses magnetic microfluidic channel and microscopic camera to characterize the magnetic moment of the cells. Through publications, it has been confirmed that this device can detect femtogram resolution change in iron concentration of biological entities. (Chalmers et al. 2017) With the advance in optical technology, we have been able to add a new feature into our system, to not only track the movement of the cells in dark-field/bright field optics, but to be able to differentiate multiple population among single sample by using different cell markers/optical filters. This was done by adding two motorized filter wheels in front of the light source and the camera, and programming/synchronizing the two filters so that simultaneous turn of both filters would give the same effect of switching between multiple optical filter cubes. This study will focus on using this system to detect a magnetic characteristic difference among group of cells, that has been shown in various literatures, such as the difference between apoptotic RBCs vs. healthy RBCs, iron-rich GBM CSCs vs. NSTCs and different sub-population of human monocytes.

To ensure that the added optical filters are working properly, the movement of rainbow beads (fluorescent calibration beads used in flow cytometry) was captured using the updated CTV system. The recorded image was then analyzed by an in-house analysis program, and results imply that the movement of the rainbow beads have been properly captured under all of the optical filters. To expand this study, the biological entities mentioned above will be magnetically characterized in CTV and the sub-groups among whole population will be analyzed further to be compared to that of the literature values. The expected result is that they portray the same magnetic characteristic and provide more information about their iron contents.

Most of the laboratory magnetometers for biological application, such as superconducting quantum interference device (SQUID), can measure the magnetic susceptibility of only one population at a time. However, if there are multiple population among the sample, pre-separation, such as FACS, density gradient or magnetic separation is required. The ability of the fluorescent CTV to magnetically differentiate multiple sub-groups using the already existing various surface markers in single sample can be used for various applications, such as detection of circulating tumor cells, anemia chronic disease, and many iron-related diseases. Also, the mechanism of the fluorescent CTV system could be developed into smaller scale, point-of-care medical devices as well, which will be investigated further in the near-future.

579324 On-chip bio-sensor for microbial contamination

Quantification of microbial assay for possible food and water contamination, especially for long-duration space missions is critically important to understand potential risk to the crew members. In this study, on-chip biosensor was fabricated for rapid screening and quantification of microbial contamination. Low-*k* dielectric (SiO_2, Gr- doped SiO_2) porous thin-film layers were deposited on ITO by surface templating assisted sol-gel method. Gels were prepared with alkoxides dispersed in ethanol containing CTAC or pluronic123 by varying concentration of precursors and surfactant, and aging time. As spun thin-film substrates were subjected initially to soft-bake followed by annealing at 400°C and characterized for porosity, pore size, and refractive index. Hydrophilic/ hydrophobic nature of the films were characterized by the contact angle measurements. Gr – SiO_2 porous layer was immobilized with the anti-microbial peptides (AMP) and specific binding ability of microbes was studied by the conductivity measurements. Fabrication and characterization of porous low-*k*(modified with Gr) thin-films, and results obtained with AMP binding and biosensing ability will be presented.

560217 Malaria Smartphone Diagnostic Using Isothermal Amplification

Title: Malaria Smartphone Point-of-Care Diagnostic Using Isothermal Amplification

Ashlee J. Colbert[1], Katherine N. Clayton[1,2], Tamara L. Kinzer-Ursem[1], Jacqueline C. Linnes[1]

[1]Weldon School of Biomedical Engineering, Purdue University, West Lafayette, Indiana 47907 [2]School of Mechanical Engineering, Purdue University, West Lafayette, IN 47907

Malaria contributes to over 400,000 deaths a year. Despite advances in technologies, developing countries are still endemic to malaria. Advanced laboratory facilities in malaria endemic countries/regions may be inaccessible, causing an increase in malaria related deaths. Furthermore, many low resource areas cannot afford the necessary infrastructure to maintain the equipment for malaria diagnoses. Indeed, there is a need for a simple, robust point of care diagnostic for physicians to accurately diagnose malaria. The current technologies for malaria detection include white light microscopy, polymerase chain reaction (PCR), and rapid diagnostic tests (RDTs). Microscopy is considered the gold standard method for detecting malaria, however the technique is time consuming, requires stained blood smears, and advanced microscopists. Microscopy is cheap and can determine the species of malaria parasites. However, developing countries do not always have access to expert microscopists and will falsely identify levels of parasites in the blood. On the other hand, rapid diagnostic tests use antibodies for detection of malaria antigens instead of DNA in blood and are easy to use. Yet, RDTs are inconsistent and have low sensitivity when parasite levels are near 100 parasites/µl. Finally, PCR methods have increased sensitivity, but are costly, time consuming, require advanced laboratory equipment, and advanced scientific training for technicians.

We have designed a loop mediated isothermal amplification (LAMP) technique for specific and robust detection of the malaria strain, *Plasmodium falciparum*. LAMP uses six primers to displace and aid in the exponential amplification of malaria DNA. The amplified blood sample is placed into a small microfluidic chip containing 400 nm streptavidin coated fluorescent beads. One of the six primers is biotinylated to bind the amplified products to the beads. This chip is placed into our smartphone device where a thirty second video is taken. The change in diffusivity of the particles in the presence of LAMP amplification products is monitored by visually tracking the Brownian motion of the beads via particle diffusometry (PD). PD diffusion coefficients decrease when Brownian motion is hindered, indicating the presence of amplified target DNA.

We have found that LAMP-PD with a smartphone device can detect as low as 100 parasites/µl (corresponding to 0.0025% parasitemia) in blood. The diagnostic device has also been tested against other infectious disease DNA and was found to be specific to *P. falciparum*, one of the deadliest strains of the malaria parasites. The PD-LAMP smartphone platform can detect malaria in an hour, making the technique promising for sensitive and robust point-of-care malaria detection.

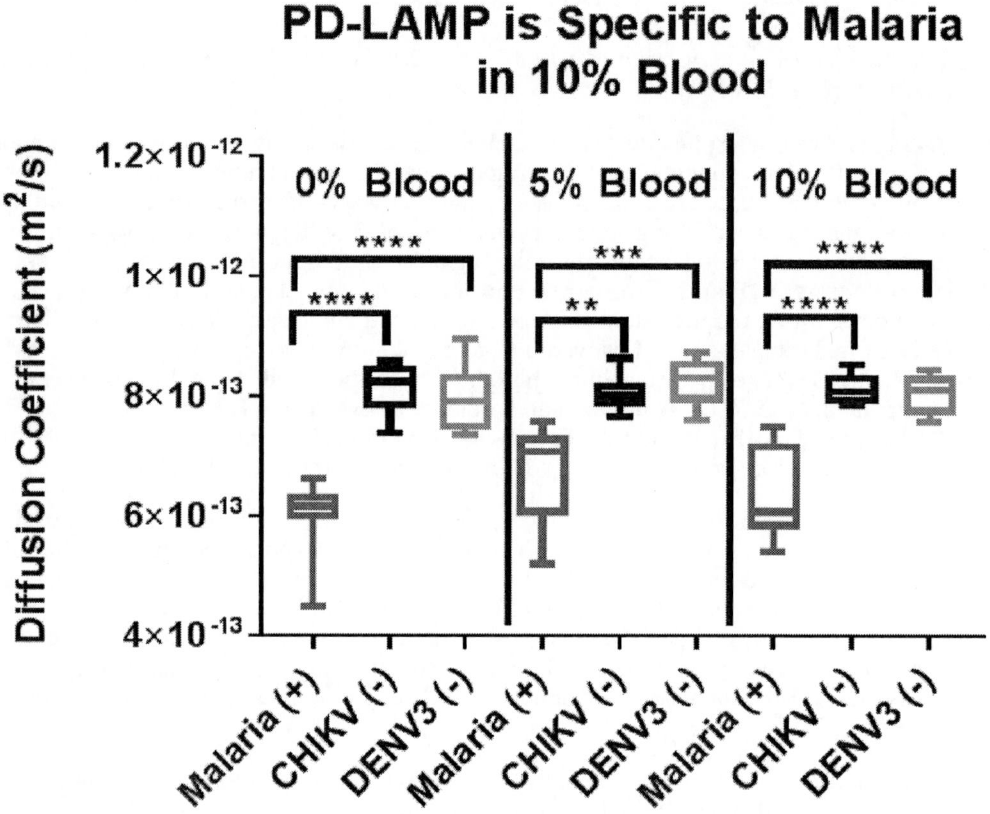

PD-LAMP is Specific to Malaria
in 10% Blood
Diffusion Coefficient (m2/s)
1.2×10-12
1×10-12
8×10-13
6×10-13
4×10-13
0% Blood
5% Blood
10% Blood
****
****
***
**
****
****
Malaria (+)
CHIKV (-)
DENV3 (-)
Malaria (+)
CHIKV (-)
DENV3 (-)
Malaria (+)
CHIKV (-)
DENV3 (-)

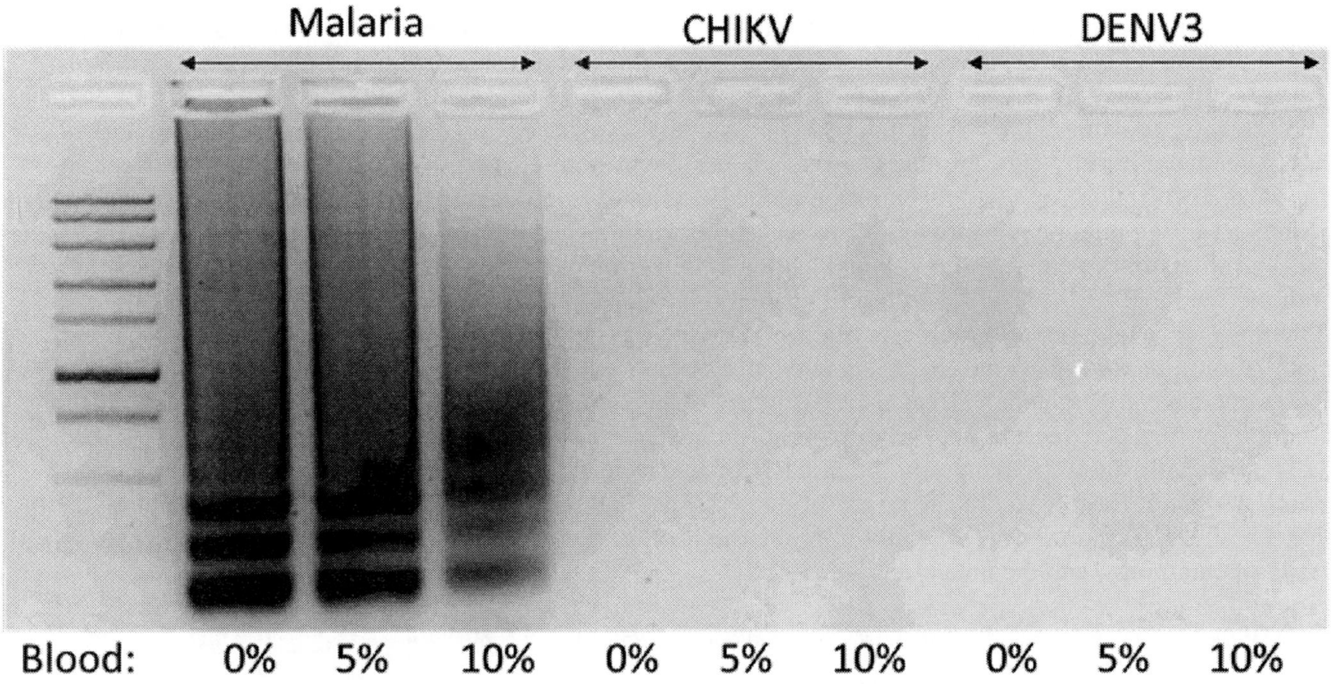

Figure 1. PD-LAMP is specific to malaria when tested against Chikungunya Virus and Dengue Virus (Type 3) at the same concentrations (top). A significant difference is shown between positive samples and controls. There is no significant difference between controls when using PD-LAMP. 2% agarose gel confirms LAMP amplicons are present only for targeted malaria DNA (bottom).

574457 Hollow Fiber Microfiltration Principles Impacting Pathogen Recovery to Enable Rapid Detection

Existing enrichment methods used for food pathogen detection often require >24 hours to produce detectable pathogen levels. There is increasing demand for methods which more quickly (<8 hours) isolate and detect pathogens from food or water samples. Previously, several continuous cell concentration devices were prototyped using hollow fiber microfilter modules. Using these prototypes, low levels of bacterial pathogens were rapidly concentrated (within 8 hours) from poultry, egg, and spinach samples. Effective use of microfiltration for cell concentration reduces the sample volume while releasing the microorganism of interest in the retentate. Improved bacterial recovery from the filter translates to greater CFU concentrations in the eluate – this permits pathogen detection at lower initial concentrations. Here, flow on both the feed and permeate sides of the filter membrane was modulated using two pumps (dual flow), and flow rates and pressures of the hollow fiber module feed, retentate, and permeate were monitored. Flow modulation directly affects the linear velocity, transmembrane pressure, and surface flow in the filter. These operational parameters affected the recovery of bacteria from a sample. When using a back-flushing process, for instance, recovery greatly improved in buffered peptone water samples – 36% versus <1% without back-flushing. In this work, the impacts of linear velocity, surface flow, transmembrane pressure (flux), and back-flushing on the recovery of bacterial pathogens are discussed. Improved bacterial recovery from the microfiltration process would improve the limit of pathogen detection across various sample types while maintaining a rapid concentration process.

580446 Physical and Chemical Approaches for Enhancing Optical Biosensors

Optical biosensors based on surface resonance plasmon (SPR) and surface-enhanced Raman scattering (SERS) have emerged as a powerful analytical and sensing technique with broad applications, such as food safety, biomedicine, homeland security, and environmental monitoring, because of the label-free, real-time detection capability offered by SPR biosensors and molecular specificity and high sensitivity enabled by SERS. For the SPR biosensors, I will cover our recent effort in the development of dually functional interdigitated electrodes to sustain SPR and to increase bacterial mass transport through external application of dielectrophoresis for the detection of low concentration bacteria. For SERS biosensors, since electromagnetic enhancement dominates the SERS effect due to extremely strong local electric fields (i.e., hot spots) induced by localized surface plasmon resonance (LSPR), the development of plasmonic nanostructures as SERS-active substrates is one of the frontiers in this field. I will first talk about our effort on the development of SERS-active plasmonic nanostructures, including quasi-3-dimension plasmonic nanostructure arrays (Q3D-PNAs) and the nanostructures enabling the extension of electric field, for sensitive detection of small and large analytes. As detection using SERS is typically carried out on bare metal surfaces, because of the near field effect, non-specific adsorption frequently occurs and significantly reduces detection performance. In this regard, I will talk about our effort on the development of stealth surface modification and hierarchical zwitterionic modification for SERS substrates to enable the sensitive and specific detection in protein solution and real-time drug monitoring from blood plasma, respectively.

577622 Invited: Wearable Fabric Sensor for Determination of Sodium Ions in Sweat

A study done on college athletes showed only 34% were properly hydrated. Failure to keep the body properly hydrated can lead to a variety of health complications, including ulcers, high blood pressure, and kidney disease. In addition to health hazards, improper hydration can significantly hinder athlete performance. The majority of endurance athletes interviewed listed hydration as one of their top concerns during a race. Currently available monitors such as Garmin, Fitbit, and the Apple Watch log miles/pace/calories burned by measuring heart rate at the wrist, and some add-on devices track heart rate. However, in order to obtain physiological information, such as dehydration or lactic acid levels, athletes must use point-of-care devices (e.g. Lactate Plus) that often require a finger prick blood draw from the athlete. The trend of monitoring fitness and physiological statistics via smart apparel and accessories has seen a steady growth over the past decade and is predicted to continue to increase.

We have developed flexible, lightweight fabric materials that can selectively determine physiological information from sweat forming on the surface of the skin. The technology is the first lightweight fabric sensor to provide real-time information regarding hydration levels during exercise or training through selective determination of sodium ion levels. As sodium ion levels in the sweat increase, athletes must replace these electrolytes to remain hydrated. These dehydration patterns are unique and vary based on diet, weather conditions, and human physiological cycles. Therefore, there is not a "one size fits all" hydration program. The ability to monitor an individual's hydration parameters is unique to the our sensor and gives it a competitive advantage over "sweat monitors" that monitor water lost or "water timers" that remind the athlete to drink water every 20 minutes.

In spite of advances made towards the detection of biomarkers in sweat, there is no sensor capable of long-term detection in constricted or load-bearing applications where other flexible plastic sensors might cause discomfort. We have developed a flexible mat sensor made of multiwall functionalized nylon-6 material. The material is also functionalized with a molecule that reacts with sodium ions. As the sensor reacts with sodium ions in sweat, a complex-molecule forms which impedes the flow of an electrical current. This change in the current flow can allow the sensor to quantify the amount of sodium present. The proposed sensor will be measure overall sodium ion concentration and changes in sodium ion concentration, so that all wearers will be able to determine if they are suffering from an electrolyte imbalance, regardless of starting point.

In this talk, we will discuss the ability of our sensor to withstand stress, strain, and bending. Uniaxial and biaxial tensile testing was performed on both neat nylon and the functionalized sensor material. Preliminary results show that the addition of nanotubes into the nylon matrix increase strength, without sacrificing stretch. Subject testing of our prototype sensor on 30-50 participants will also be discussed.

576385 Understanding How Pollution Episodes Affect Community-Level Air Quality with a Distributed Sensor Network

Short- and long-term exposure to fine particulate matter (PM2.5) pollution is linked to numerous adverse health effects, and acute events, like wildfires and fireworks, can cause dramatic increases in PM2.5 levels. Although fewer studies have examined the health effects of PM2.5 from these types of events, several studies suggest wildfire smoke and fireworks cause adverse respiratory effects. Pollution impacts from wildfires are becoming an increasing concern as both the number and size of wildfires continues to increase. In fact, although air quality has improved in the US over the past 30 years, it has declined in wildfire-prone states. Conventional PM2.5 measurements are expensive, resulting in spatially and temporally sparse data that may not accurately represent local-scale pollution gradients and may be slow to detect a pollution event. New low-cost PM sensor technologies offer the potential to improve the spatial and temporal resolution of PM2.5 measurements although they have limitations in terms of accuracy and precision. Here we demonstrate how air quality measurements from a network of 134 citizen-hosted, low-cost PM2.5 sensors can be integrated with dynamic data-fusion algorithms and visualization strategies to understand and communicate highly resolved PM2.5 concentrations in an intuitive way during two pollution events in the Salt Lake Valley. The results illustrate how local policy on fireworks affects community-scale PM concentrations and how the network can serve an early warning of an approaching pollution event.

Drs. K. Kelly and P.-E. Gaillardon have an interest in the company Tetrad: Sensor Network Solutions, LLC, which commercializes solutions for environmental monitoring.

576010 Novel Chromogenic Sensors Enabled By Multi-Stimuli-Responsive Shape Memory Polymers

Here we report novel chromogenic sensors based on smart shape memory polymers (SMPs) exhibiting nontraditional all-room-temperature shape memory (SM) effects. Stepwise recovery of the collapsed macropores with 350 nm diameter created by a "cold" programming process leads to easily perceived color changes that can be correlated with the concentrations of swelling analytes in complex, multicomponent non-swelling mixtures. High sensitivity (as low as 10 ppm) and unprecedented measurement range (from 10 ppm to 30 vol.%) for analyzing ethanol in octane and gasoline have been demonstrated by leveraging colorimetric sensing in both liquid and gas phases. Proof-of-concept tests for specifically detecting ethanol in consumer medical and healthcare products have also been demonstrated. These sensors are inexpensive, reusable, durable, and readily deployable with mobile platforms for quantitative analysis. Additionally, theoretical modeling of solvent diffusion in macroporous SMPs provides fundamental insights into the mechanisms of nanoscopic SM recovery, which is a topic that has received little examination. These novel sensors are of great technological importance in a wide spectrum of applications ranging from environmental monitoring and workplace hazard identification to threat detection and process/product control in chemical, petroleum, and pharmaceutical industries.

575491 Real-Time Monitoring of Supercritical Hydrocarbon Fuels

Jet fuel is used in high-performance aircraft as the primary heat sink, which causes chemical changes inflight that alters fuel properties and can affect flight performance. Trace species in the fuel can drastically change how the fuel reacts to the increased temperature and lead to deposition of solids, resulting in increased maintenance of aircraft components. As new, high-performance aircraft are developed and fuel is taken to higher temperatures (160 – 500 °C), real-time monitoring of fuel during flight and during system development is needed to better understand the fuel degradation process and enable reliable aircraft operation. At temperature and pressure, tracking of the trace components that cause bulk fuel degradation is challenging due to low concentrations and a constantly changing matrix. Mainstream Engineering has continued development of a Raman spectroscopic sensor that uses a fiber-coupled, in-line, flow cell capable of withstanding 550 °C and 3.8 MPa to measure fuel composition in real-time. The flow cell was used to obtain high-quality spectra of different fuels from room temperature up to 550 °C and calibrations were used to identify changing concentrations of fuel components. The system has been used to examine fuel degradation in the thermal oxidation regime (160 – 350 °C) and in supercritical fuel in the pyrolytic regime (350 – 550 °C). Method development and sensor performance from room temperature to 550 °C as validated with gas chromatography/mass spectrometry and 3D fluorescence spectroscopy will be presented.

565603 Polymer Thin Film Platforms for the Rapid Detection and Analysis of Waterborne Radionuclides

The detection and isotopic analysis of uranium and plutonium special nuclear materials (SNM) in environmental waters is critical for nuclear forensics and nonproliferation efforts. Techniques for isolating and concentrating SNM for isotopic analysis are time-consuming and often have to be performed offsite. The ability to perform rapid, onsite analysis of waterborne SNM would be a powerful nuclear forensics tool; currently no such technology exists.

This presentation describes the use of SNM-selective polymer thin films for the isolation, detection, and isotopic analysis of waterborne SNM via alpha spectroscopy. The films were coated onto silicon wafers and porous membranes through spin coating. This method of film casting provided a high-degree of control over the thickness and surface uniformity. Several polymer film chemistries were compared for their ability to capture uranium and plutonium from water at low and circumneutral pH, and isotope peak resolution were determined for films with thicknesses from 25 to 250 nm. The films afforded high isotope peak resolutions, given as full width at half maximum, that were measured to be 30-60 keV. Application of the films as an SNM selective layer on the surface of macroporous membrane supports enabled the flow-through processing of large sample volumes and concentration of low-level SNM at the membrane surface for direct spectroscopic analysis. Results of this work are being used to develop a field-portable device for rapid detection and analysis of waterborne radionuclides.

572090 Real-Time Membrane Permeability Measurements of Hydrogen Sulfide and Acetone

The permeability of membranes plays a critical role in many physiological as well as industrially relevant processes. Significant laboratory data already exist on the permeability of a number of molecules in a variety of membranes. However, rapid on-site measurement of permeability in real-time is of interest because: (1) membrane characteristics change during processing and use and (2) the need for quality control during high throughput engineering. Development of synthetic membranes is a rapidly growing area of research and development, driven by the variety of biomedical applications, from skin substitutes to organ tissue.

Among the molecules of interest in physiological processes are hydrogen sulfide, a hydrophobic molecule, and acetone, a highly water-soluble molecule. We will here present membrane permeability measurements of these molecules obtained utilizing an innovative combination of electrocatalytic and colorimetric methods that is amenable to miniaturization, portability, and the ability to measure at low parts-per-billion levels. Measurements have been performed in both synthetic membranes and animal skin. The figure below is an example of measurements made of permeation rate of hydrogen sulfide through a 25 mm thick synthetic membrane having very low permeability. Details of the measurement systems and apparatus as well as results of the measurements will be presented. The authors will also discuss how the observations can be applied to several biomedical and analytical approaches.

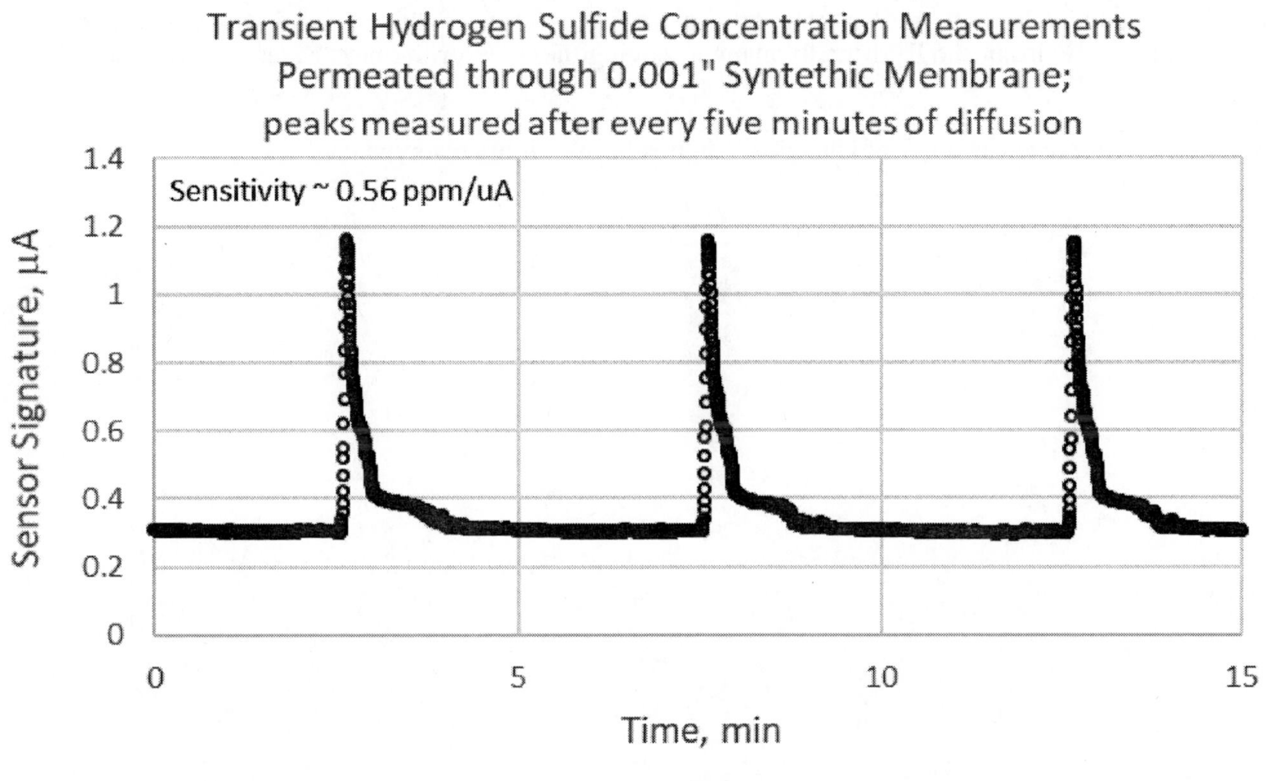

563880 Ultra-Stretchable Conductive Polymer Complex As a Wearable Strain Sensor with Excellent Linearity and Repeatable Autonomous Self-Healing Ability

Wearable strain sensors are essential for the realization of applications in the broad fields of remote healthcare monitoring, soft robots, immersive gaming, among many others. These flexible sensors should be comfortably adhered to skin and capable of monitoring human motions with high accuracy, as well as exhibiting excellent durability. However, it is challenging to develop electronic materials that possess the properties of skin—compliant, elastic, stretchable, and self-healable. This work demonstrates a new regenerative polymer complex composed of poly(2-acrylamido-2-methyl-1-propanesulfonic acid) (PAAMPSA), polyaniline (PANI) and phytic acid (PA) as a skin-like electronic material. It exhibits ultrahigh stretchability (1935%), excellent repeatable autonomous self-healing ability (repeating healing efficiency > 98%), and exceptional linearity ($R^2 > 0.995$) — outperforming current reported wearable strain sensors. The deprotonated polyelectrolyte, multivalent anion, and doped conductive polymer, under ambient conditions, synergistically construct a regenerative dynamic network of polymer complex crosslinked by hydrogen bonds and electrostatic interactions, which enables ultrahigh stretchability and repeatable self-healing. Sensitive strain-responsive geometric and piezoresistive mechanisms of the material owing to the homogenous and viscoelastic nature provide excellent linear responses to omnidirectional tensile strain and bending deformations. Furthermore, this material is scalable and simple to process in an environmentally-friendly manner, paving the way for the next generation flexible electronics.

579268 A Review and Outlook on Machine Learning and Data Science in Chemical Engineering

With today's development of the internet, internet of things, wireless sensors and communications, smart manufacturing, and high-throughput experiments, the amount of data collected has grown exponentially, analogous to the Moore's law in the increase of solid-state transistor density. The data explosion has made all sectors including engineering, medicine, business, commerce, finance, and materials science to endorse the potential and power of big data. The relevance of big data to chemical engineering and science is evident from the recent technical contributions by chemical engineers and scientists.

Great successes in machine learning today have to do with natural language processing (NLP), multi-lingual translation, image recognition, etc. In addition, the demonstration of AlphaGo by Google's DeepMind in early 2016 was one of the most shocking events to the world. However, the game of Go has little in common with the degree of uncertainty and the complexity of real-world engineering problems. In this talk we discuss the characteristics of process systems engineering problems and provide a brief review of machine learning and data analytics in chemical engineering in the last three decades. We give a few attributes for machine learning techniques to be widely applicable in the process systems. Owing to these characteristics, process monitoring and inferential sensors are two of the most adopted data-driven applications in chemical engineering.

The talk will then provide an assessment of the resurgent interest in deep neural networks and deep reinforcement learning and contrast them against the technology two decades ago. We also give a few examples of high-profile applications that failed the test of time. We assert that most of the sustainable applications are domain-specific, in that they incorporate specific structures and knowledge in the models. We will briefly touch on the tremendous advancement in statistical machine learning, which are the core of data science and could propel the advances in other scientific and engineering disciplines, such as smart manufacturing, bioinformatics and materials science.

559766 Neural Network-Based Modeling and Operation for ALD of SiO_2 Thin Films Using Data from a Multiscale CFD Simulation

Atomic layer deposition (ALD) plays an important role in the manufacturing of highly conformal thin-film materials, which are used extensively in the semiconductor industry. In our previous study, a multiscale computational fluid dynamic (CFD) model has been developed for the SiO2 ALD process in an optimized showerhead ALD reactor geometry. The optimized reactor design involves a volume adjusting horn in the upstream region and a radially adaptive sizing of the showerhead holes [1], [2] . The results from the developed macroscopic CFD model show that those two design features enable quick achievement of steady-state in the gas-phase domain of the reactor, and provide a relatively uniform surface pressure and temperature profile for the microscopic kinetic Monte-Carlo (kMC) modeling of the surface deposition, which authentically simulates the film deposition process [3].

However, the multiscale CFD model is computationally demanding and is not suitable for applying real-time control schemes like model-based controllers (MPC). Motivated by this, in this work, we investigate the transient operation of the full reactor in details, using the model developed previously. The impact of operation parameters, including inlet process gas flowrate, temperature and pressure, on the deposition process is discussed and a corresponding database with temporal information of the process is constructed. Also, the appropriate operating boundary accounting for both microscopic and macroscopic operations is determined to ensure process feasibility [3]. Subsequently, a recurrent neural network (RNN) is constructed and trained using the model-generated database, which enables the prediction on the transient process dynamics based on the inputs of operating condition [4]. Moreover, utilizing the RNN model, a control scheme is designed based on the developed RNN to reduce the impact of potential disturbances on the deposition process and enhance the film uniformity. Finally, an operation example is shown to demonstrate the practicability of the proposed method.

[1] Lee, C.S., Oh, M.S., Park, H.S. US Patent No. US7138336B2. 2003. Retrieved from

https://patentimages.storage.googleapis.com/01/e8/3e/5c751ffe023e4d/US7138336.pdf

[2] Crose, M., Zhang, W., Tran, A., Christofides, P.D. Multiscale three-dimensional CFD modeling for PECVD of amorphous silicon thin films. Computers & Chemical Engineering. 2018, 113, 184 – 195.

[3] Ding, Y., Zhang, Y., Kim, K., Tran, A., Wu, Z., Christofides, P.D. Microscopic Modeling and Optimal Operation of Thermal Atomic Layer Deposition. Chemical Engineering Research and Design. 145, 159-172, 2019.

[4] Connor, J.T., Martin, R.D., Atlas, L.E. Recurrent neural networks and robust time series prediction. IEEE Transactions on Neural Networks. 1994, 5, 240-254.

579520 Keynote Talk: Intelligent Cloud-Based Algorithms for Pharmaceutical Drug Development

The ability to predict or explore polymorphism in organic crystal compounds is of great importance for many industries since many physical and chemical properties of molecular crystals strongly depend on the molecular packing in the solid state. Crystal structure prediction (CSP) aims to reduce the risk and improve the efficiency of pharmaceutical drug manufacturing by correctly predicting the low energy crystal forms of organic molecules from minimal information, such as the chemical diagram. At XtalPi, we have developed a cloud-based computational platform that combines advanced physics-based algorithms with A.I/machine learning algorithms. Our CSP platform can rapidly generate and rank millions of crystal structures for pharmaceutically relevant systems and we highlight some cases in which our CSP platform has revealed low energy polymorphs in challenging systems. We will also discuss some areas where we are expanding our computational platform to make accurate predictions on the physiochemical and pharmaceutical properties of small-molecule candidates for drug design, solid-form selection, and for other critical aspects of drug development.

568479 Industrial Success Stories with Machine Learning and Deep Learning in Matlab

Machine learning has rapidly gained in popularity, spurred by well-publicized advances like deep learning and widespread commercial interest in big data analytics. Despite the enthusiasm, companies and researchers are struggling to deploy machine learning models in chemical plants so as to reconcile the hype with actual application and success.

In this presentation, MathWorks will showcase the successes of machine learning and deep learning in industrial settings with 6 user stories/case studies - all drawn from the process industries. With these examples drawn from customers' usage of MATLAB, MathWorks seeks to dispel the notion that machine learning is solely the subject of academic study but can indeed be applied to process plants with measurable business benefits.

In addition, machine learning workflow elements will be discussed in this talk including pre-processing, model building using apps and deployment to industrial systems using MATLAB.

569189 Keynote Talk: Modern Data Analytics to Accelerate the Design of Advanced Materials

We will discuss three critical aspects of applying a machine learning approach to accelerate the design of advanced materials for energy applications: 1) consistently measured, relevant experimental data with known pedigree, 2) in-depth scientific features that can be correlated with target properties, and 3) training of surrogate machine learning models. We will present examples of predicting mechanical properties and oxidation response of high-temperature alloys, thermochemistry of complex oxides, and melt-pool microstructure of single-track by the additive manufacturing process.

The research was sponsored by the LDRD Program of Oak Ridge National Laboratory and the Department of Energy, Vehicle Technologies Office, Propulsion Materials Program.

570365 GPU-Accelerated Computation of Stirred Tanks

A fast and economical solver, accelerated by the Graphics Process Units (GPU) of a single graphics card in a desktop computer, is developed for the simulation of stirred tanks, integrating Lattice Boltzmann Method (LBM), coherent Large Eddy Simulation (LES) and Immersed Boundary Method (IBM). The grid resolution can reach 13.8 million nodes in maximum, resolving considerable details of flow field in stirred tanks with only such a simple desktop computer. In the meantime, the computational speed is 1500-fold faster than that of the traditional transient simulation of Computational Fluid Dynamics (CFD) based on Navier-Stokes equations implemented on 16 CPU cores. It only takes less than two minutes to update the simulation of one impeller revolution, enabling the transient simulation of longer physical time of stirred tanks. We find that at least 4,000 impeller revolutions in simulation is needed to achieve an obviously time-independent macro-instability frequency. With the coherent LES model and reasonable grid resolution, the fast solver is able to resolve more than 95% of total Turbulent Kinetic Energy (TKE) for highly turbulent region, reproducing the monotonic decrease of TKE along the radial direction, much better than the Smagorinsky model. The average velocity and TKE predicted by the fast solver are in better agreement with the experimental data in literature. The solver is therefore more suitable for the fast simulation of stirred tanks using only a desktop computer without the need of much finer grid resolution and supercomputers. We believe this high-resolution approach serves as a good gateway to further integrate with Artificial Intelligence for chemical reactors.

564376 Keynote Talk: Faster Prediction with Artificial Intelligence: Using Process Simulations and Production Data to Develop Fast-Running Inference Models for Manufacturing

Navigating a large-dimensional design space with simulations is a significant improvement over doing so with physical experiments but can still be very time consuming. Deep learning methods are often used to develop fast-running surrogate models of more computationally expensive simulations in order to get near-real-time predictions and to optimize complex systems such as manufacturing processes, energy systems, and fusion reactors. In this talk, I will discuss an application of machine learning to develop a fast-running surrogate models that captures the dynamics of industrial multiphase fluid flows. I will also discuss an improved population search method that can help the analyst explore a high-dimensional parameter space to optimize production while reducing the model uncertainty. Results from demonstration problems and from real-world Inertial Confinement Fusion simulations will be presented.

575905 (Invited Plenary Talk): Reproducibility and Reusability of Surface Bound Electrochemical Sensors

Polymer structures on sensor surfaces are known to have reproducibility and reusability issues caused by hydrodynamic shifts, minor changes in surface coverage, charging of the surface, and steric restructuring of the polymer surfaces. Within our laboratory, we primarily use elastin-like polymers and hydrophobic-self-assembled-monolayer:cyclodextrin complexes as functional polymers on sensor surfaces. Often, with these polymers, the same methodology does not lead to the same surface modification due to small minor surface changes including physiosorbed surface content or steric hinderances. Therefore, attention and strategy are needed to ensure a reproducible sensor surface. Further, we use electrochemical impedance spectroscopy, a highly sensitive but difficult to reproduce strategy in biosensors because small surface changes can result in large changes in the recorded signal. We will discuss our methodological approaches to achieve reproducible and reusable sensor using polymer surfaces with electrochemical impedance spectroscopy. Our group cares about reusability as a long-term sensor strategy towards measuring bioanalyte content *ex vivo* in biofluid samples. We will discuss both positive and negative approaches to generate the same results over multiple sensors, and the ability to regenerate a sensor for reuse. Further, we will talk about voltage dependency of the surface and ability to reproducibly detect nanomolar chemical changes.

The author would like to acknowledge the support from NSF EAGER (CBET 1638896) and NIH CIBBR (P20GM113131). This contribution would not be possible without the endless support of the undergraduate students, graduate students, post doctorates, and collaborators of the Surface Enhanced Electrochemical Diagnostic Sensors Laboratory.

576041 (Invited Plenary Talk): Electrochemical Sensors for Multiplexed Infection Detection and Monitoring

Bacterial cells secrete hundreds of different molecules as part of their natural processes. These molecules range from quorum sensing molecules or pheromones for communication, to siderophores as nutrient scavengers, to virulence factors that attack intruders. The ability to quantitatively measure the concentrations of these molecules is important for both research and clinical applications. This talk will describe electrochemical sensors that are being developed and used for diagnosing infection and monitoring bacterial physiology. The sensors utilize voltammetry to measure interactions of redox-active molecules with electrode surfaces as a signal transduction mechanism to track binding of target molecules with recognition elements. The method works directly in biological fluids, allowing ex vivo measurements to be easily performed.

569320 (Invited Plenary Talk): Hybrid Plasmonic Nanomaterials for Uranium Sensing

Uranium, a radioactive material with a long half-life, accumulates in the environment in its oxidative form uranyl, which can contaminate soil and water. Importantly, uranyl forms complexes with anions and cations thereby influencing solubility, toxicity, and fate of these heavy metal species. Furthermore, uranyl speciation varies with pH. The resulting complex speciation complicates detection and/or requires significant sample pretreatment. As such, methods that are capable of identifying trace uranyl species in complex samples are needed. In this presentation, hybrid plasmonic nanomaterials and experimental workflow implementation will be used to establish a rigorous protocol for uranyl sensing and detection. Namely, localized surface plasmon resonance spectroscopy (LSPR), Raman spectroscopy, and surface enhanced Raman scattering (SERS) serve as label-free and near real-time methods for identifying uranium species in complex aqueous solutions. Reproducible detection of uranyl, an important biological and environmental contaminant, from complex matrices is successfully achieved using hybrid plasmonic nanoparticles. Traditionally, non-specific binding of interfering species limits detects of molecules such as uranyl. Herein, this is overcome using materials design and rigorous sample analysis workflow design. Synergistic approaches for uranyl isolation and SERS detection is promising for real-world sample detection and eliminates the need of radioactive tracers and extensive sample pretreatment steps.

579768 (Invited Plenary Talk): Continuous, Real-Time, Physiological Monitoring with Nanosensors

Continuous physiologic and therapeutic monitoring has the potential to redefine prediction, diagnosis, and treatment of disease. Current approaches focus on intermittent blood sampling and analysis – an approach that is not scalable or predictive with decisions based on cohort values rather than individual history. Future approaches will draw on continuous monitoring to track metabolic parameters before, during, and after treatment. This enables real time feedback, as well as comparison to a patient's own historical trends and fluctuations. Our nanosensor platform has the ability to continuously monitor analyte levels with straightforward optical measurements. We, and others, have developed sensors for a wide variety of ionic analytes (e.g. sodium, potassium, lithium, pH) as well as nonionic analytes (e.g. glucose, histamine) which are relevant for monitoring in clinical and research settings. In this presentation I will discuss design and fabrication of nanosensors, control of sensor response, incorporation of next generation fluorescent tools into our sensors, and applications of this platform for physiological monitoring and the study of biofilm metabolism. This sensor approach shows potential as a research or medical tool for continuous monitoring of drug and metabolite concentrations without the need for direct blood samples.

578760 Photoelectrochemical Sensor

The evolution of photoelectrochemical (PEC) bioanalysis has resulted in substantial progress in its analytical performance and biodetection applications. PEC sensor represents a unique means for chemical and biological detection, with foci of optimizing semiconductor composition and electronic structures, surface functionalization layers, and chemical detection methods. Here, I have briefly discussed my recent developments of nanowire-based PEC sensing, with particular emphasis on three main detection mechanisms and corresponding examples. I have also demonstrated the use of the PEC sensing of real-time molecular reaction kinetic measurements, as well as direct interfacing of living cells and probing of cellular functions. My work will serve as a useful source to inform the interested audience of the latest developments and applications in the field of PEC bioanalysis.

580152 High-Throughput Sensing of Single-Bacterium Growth: Toward Rapid Antibiotic Susceptibility Testing

Accurately measuring the number of viable microorganisms plays an essential role in microbiological studies including antibiotic susceptibility testing (AST). Since conventional culture-based AST methods are too time-consuming (typically 24-72 h), rapid AST is urgently needed for preventing the increasing emergence and spread of antibiotic resistant infections. Although several phenotypic antibiotic resistance sensing modalities are able to reduce the AST time to a few hours or less, concerning the biological heterogeneity, their accuracy or limit of detection are hampered by low throughput. Here, we present a rapid AST method based on whole slide imaging (WSI)-enabled high-throughput sensing of single-bacterium growth, conducted on custom-built on-glass-slide culturing device. The time for determining the minimum inhibitory concentration (MIC) was theoretically shortest that ensures the growth of each individual cell present in a large population is inhibited. Reliable MIC of *E. cloacae* against gentamicin was obtained within 1 h, while the gold standard broth dilution method required at least 16 h for the same result. Owing to the feature of high-throughput quantitative analysis, our approach can be further applied to timely identifying the rising antibiotic resistance adapted to the treatment and guiding the adjustment of treatment strategy to prevent the aggravation of antibiotic resistance.

570379 Ionophore-Based Optical Sensors Incorporating Long-Lifetime Phosphorescent Microparticles for Background-Free Ion Detection in Biological Samples

Fluorescent sensors are ideal for biological applications due to their low invasiveness and miniaturizability. However, signal interference from autofluorescence and scattering limits their practical application. Long-lifetime phosphorescent materials have emission lifetimes on the order of hours or even days, and have long been used commercial in street signs and 'glow-in-the-dark' materials. More recently these materials have attracted interest as agents for biological imaging and sensing as a method for reducing background noise, since, by programming a delay between excitation and emission collection, the short-lived signals from scattering and autofluorescence can be avoided completely. In this work we demonstrate an ionophore-based optical sensor incorporating long-lifetime phosphorescent microparticles for background-free sodium detection in biological samples. We couple the phosphorescent microparticles to a pH-sensitive quencher dye and sodium-binding ionophore to create a dynamic and reversible optode film selective for sodium. We also characterize the sensor in terms of selectivity and stability and demonstrate its usefulness to detect sodium levels in human serum.

576752 Quantification of Biomarkers at Point of Need Using Metal Enhanced Fluorescence and Surface Acoustic Waves

Point of need (PON) quantification of cancer biomarkers such as those used in monitoring disease progression requires sensing at low (pg/ml) limits of detection from body fluids such as blood. For example, proteins such as CEA and CA19-9 are monitored by clinicians for progression of pancreatic cancer, and CA125 is monitored for ovarian cancer progression. Proteins such as PSA are used for initial screening for prostate cancer. All such tests are currently performed in clinical laboratories, off site. The capability to provide such monitoring at the clinician's office or hospital is challenging, as these proteins need quantification at low, pg/ml to ng/ml levels from a body fluid such as blood, which contains interfering proteins at orders of magnitude higher concentrations. We present concepts and results for such detection using plasmonic (metal) enhancement of fluorescence for lowering detection limits, and Rayleigh surface acoustic waves (SAWs) for reducing interference from non-specific binding (NSB) of other proteins in an immunofluorescence format. These same Rayleigh SAWs are utilized in our platform to reduce incubation time, by inducing sample mixing.

In metal enhanced fluorescence (MEF), fluorescence intensity is increased by metallic nanostructures in the proximity of the fluorophore, leading to lower limits of detection. Recent methods for creating metallic nanostructures, such as e-beam nanolithography, colloidal lithography, and colloidal self-assembling, require complicated processes and have various drawbacks. In this work, we have developed a simple nanostructure fabrication process combining thermal annealing and oxide deposition. A thin silver film is treated with rapid thermal annealing to generate the nanostructures on a SAW device surface. These nanostructures are then coated with a thin silica film to protect them and also to control the distance between the metallic structures and fluorophores. This process provides a solution to achieving silver nanostructures over a large area with simple processing techniques. Immunofluorescence experiments conducted using these surfaces show significant enhancement of fluorescence intensity. The effect of initial silver film, and oxide coating thicknesses on fluorescence enhancement was studied in detail. All fabricated nanostructures showed enhancement, with the silica layer producing an additional enhancement. The optimized structure produced a significant intensity enhancement of ~19X, leading to the construction of an immuno-sensor with a limit of detection of a few 10's of pg/ml. Finite difference time domain (FDTD) simulations for the enhancement factor allowed for understanding the role of the silica coating, and optimization of the immuno-sensor.

Acoustic streaming from Rayleigh SAWs is utilized to remove NSB proteins to improve sensor signal in our PON sensor platform. We show that NSB proteins which interfere with sensor signal are removed effectively, and incubation times are reduced significantly utilizing these SAWs. The sensor electronics and optics for this proposed sensor platform can be easily reduced to a small, hand-held package, which we present, with further reduction in size possible. The platform is demonstrated with quantification of a typical biomarker, CEA, from blood samples.

586770 Chemically & Electrochemically Interactive Characterization of Biomaterials *in vitro*

The ever growing demand for advanced and high performance implantable biomedical materials entails development of better interactive in vitro characterization tools to obtain a more representative prediction of their performance in vivo. Herein cases of implantable biomaterials, for temporary bone scaffolding support and coronary stents applications have been discussed.

Temporary implants pose a great challenge where the implant is expected to degrade and dissolve after a period of time at which point it is either absorbed by the body or excreted in the urine. The two specific medical applications are the temporary scaffolding for bone fixation and the coronary stents application. This is to prevent the long term complications associated with presence of a foreign material in the body, for example inflammatory reactions, thrombosis and restenosis in stent applications associated with the permanent stainless steel stents. Research in this area has revolved around magnesium in the past decade as it is safely metabolized by human body upon degradation and yet possesses adequate mechanical and load-bearing properties to support a healing bone as a scaffold or an artery from elastic recoil after balloon angioplasty. The main shortcoming however is the rapid degradation of magnesium in physiological environment that causes premature failure of the implant. In a series of studies [1–7] magnesium alloys were designed and fabricated, the underlying degradation mechanism in simulated biological environment was unraveled and surface treatments were applied to fine-tune degradation rate and biocompatibility. Scanning electrochemical microscopy was extensively used as a uniquely interactive characterization technique to examine degradation of magnesium at high spatial resolution in-situ. A new bio-sensing method was introduced for detection of H2 which enabled detection of localized degradation. Localized degradation is one of the main reasons for the premature failure of load-bearing structures. The new insight into the degradation mechanism allowed for an informed exploration of surface treatments to fine-tune degradation rate of magnesium in a physiological environment. In particular, ionic liquid surface treatment, conversion coatings based on rare earth metal and a biodegradable conducting polymer coating were investigated in order to suppress the initial rapid degradation, alkalization and H2 evolution while in the longer term creating a self-healing surface film capable of maintaining uniform degradation pattern. Results of these works introduced candidates with optimized bio-degradation rate that would minimize the risk of early stage alkalization and inflammatory reaction for the surrounding tissue as well as minimizing risk of premature failure to ensure adequate mechanical support is provided during the healing (of the bone) or remodeling (of the artery).

In the field of biomaterials sciences, although there are standards developed to examine the performance of materials in certain in vitro environments, there is still major need for a testing procedure with holistic approach to take into account most environmental parameters, e.g. load and cycles, ions/electrolyte, proteins, cell adhesion, mobility, temperature etc. Currently the in vitro test can be considered only as a preliminary screening test with limited ability to give the realistic picture, and a holistic simulation procedure has to be developed to measure performance in the biological condition. For example a hip and knee simulators with the tribological test properties of the materials together with full consideration of the corrosivity in the cell culture environment and other interacting elements. It is only with a holistic approach and using a multi-array of sensors that the full impact of an implant material on the host can be determined. Surface modifications are often performed on the biomedical implants to improve corrosion resistance, wear resistance, surface texture and biocompatibility. The interactions that take place at the atomic level between the surface of the implant, the host and the biological environment including all types of micromotions of the implants similar to what is expected in the human system should be studied thoroughly in order to advance implant materials with predictable function in the body.

References

[1] S.S. Jamali, S.E. Moulton, D.E. Tallman, M. Forsyth, J. Weber, G.G. Wallace, Applications of scanning electrochemical microscopy (SECM) for local characterization of AZ31 surface during corrosion in a buffered media, Corros. Sci. 86 (2014) 93–100.

[2] S.S. Jamali, S.E. Moulton, D.E. Tallman, M. Forsyth, J. Weber, G.G. Wallace, Evaluating the corrosion behaviour of Magnesium alloy in simulated biological fluid by using SECM to detect hydrogen evolution, Electrochim. Acta. 152 (2015) 294–301.
[3] S.S. Jamali, S.E. Moulton, D.E. Tallman, M. Forsyth, J. Weber, A.S. Mirabedini, et al., Corrosion protection afforded by praseodymium conversion film on Mg alloy AZNd in simulated biological fluid studied by scanning electrochemical microscopy, J. Electroanal. Chem. 739 (2015) 211–217.
[4] Y. Zhang, X. Liu, S.S. Jamali, B.R.W. Hinton, S.E. Moulton, G.G. Wallace, et al., The effect of treatment time on the ionic liquid surface film formation: Promising surface coating for Mg alloy AZ31, Surf. Coatings Technol. 296 (2016) 192–202.
[5] S.S. Jamali, S.E. Moulton, D.E. Tallman, Y. Zhao, J. Weber, G.G. Wallace, Self-healing characteristic of praseodymium conversion coating on AZNd Mg alloy studied by scanning electrochemical microscopy, Electrochem. Commun. 76 (2017) 6–9.
[6] D. Liu, G. Xu, S.S. Jamali, Y. Zhao, M. Chen, Fabrication of biodegradable HA/Mg-Zn-Ca composites and the impact of heterogeneous microstructure on mechanical properties, in vitro degradation and cytocompatibility, Bioelectrochemistry. 129 (2019) 106–115.
[7] S.S. Jamali, S.E. Moulton, Y. Zhao, S. Gambhir, M. Forsyth, G. Gordon, A biodegradable conducting polymer coating to mitigate early stage degradation of magnesium in simulated biological fluid : an electrochemical mechanistic study, ChemElectroChem. 6 (2019) 4893–4901.

578352 Solvent-Resistant Electrochemical Transistors As Dual Sensing Platforms in Aqueous and Organic Media

Organic electrochemical transistors (OECTs) are promising contemporary sensing platforms for high-fidelity transduction of (bio)chemical signals into electrical observables. In an OECT, such signals cause can ions to inject from an electrolyte into an organic semiconducting polymer channel, changing its bulk conductivity and the output drain current. Such devices provide multiple advantages that make them apt for bioelectronics applications, namely (1) operating voltages less than 1 V, (2) direct interfacing with biologically relevant electrolytes for *in vivo* application, and (3) printable fabrication, for flexible and stretchable form factors. Despite such advantages, OECTs have only achieved high signal transduction with a narrow subset of channel materials, many of which require cytotoxic cross-linkers to prevent dissolution in aqueous environments and/or exhibit poor processability in carrier solvents. Consequently, there is a strong need to develop biocompatible channel materials with high-processability.

In this work, I discuss the development of single-component OECTs which are insoluble to both aqueous and organic media, yet can be facilely solution-processed from commercial water-soluble carboxylated polythiophene (CPTs) precursors post-synthesis. Such OECTs demonstrate robust stability and high pseudocapacitances even after repeated 2-V width voltage sweeps, as well as high transduction (transconductances = 30 mS at V_{gate} = -0.5 V) in aqueous and nonaqueous systems. Changes in doping state and ion injection will also be probed using spectroelectrochemistry and electrochemical quartz crystal microbalance to further understand the origin of high transconductance in these materials. In conjunction with other electrochemical techniques, we should be able to elucidate detailed characteristics that dictate high mixed ionic-electronic transport in these materials.

563640 Investigation of Polymer-Plasticizer Blends for BTEX Sensing Using Acoustic Wave Devices

Volatile organic compounds (VOCs) are chemicals which evaporate easily and become gases at ambient temperature and pressure. The EPA has made the monitoring of VOCs mandatory due to their health impacts, which range from headaches and nausea to cancer. The current techniques used for VOC monitoring such as photo ionization, gas chromatography, mass spectroscopy, and e-noses sensors are expensive, time-intensive, demand rigorous sample preparation and cannot quantify the VOCs in the air. To employ the shortcomings of current sensor systems, quartz crystal microbalance (QCM) and surface acoustic wave (SAW) sensors based on polymer-plasticizer thin films have been developed. Acoustic wave devices are capable of detecting very low concentrations (ppm, ppb) of organic compounds in the environment. Imprinted polymers as sensing film have been used to detect BTEX gas analytes to a lower limit of 0.5ppm using SAW devices.

Traditionally, the sensitivity and performance of rubbery polymers towards hydrocarbon sensing are superior to glassy polymers however, they show low molecular selectivity due to their amorphous nature, availability of large free volume elements, and display of undesirable viscoelastic effects. Plasticization helps to modify the chemical and physical properties of the glassy polymer by decreasing the glass transition temperature, altering the pore dimensions and increasing the free volume of the polymer film thus enabling higher diffusion and sorption of the analyte molecules. This work shows that the selectivity and sensitivity of glassy polymers can be tailored by introducing a plasticizer. Poly (ethyl methacrylate) (PEMA), Poly (methyl methacrylate) (PMMA), PEMA-PMMA, PMMA-Polystyrene copolymers were modified by introducing plasticizers, diisononyl cyclohexane-1, 2-dicarboxylate (DINCH), diisooctyl azelate (DIOA), dibutyl sebacate (DBS), n-butyl stearate (BS), and di-n-butyl phthalate (DBP) to enhance BTEX sensitivity and selectivity in presence of interferants.

The objective of this study is to investigate the effect of various plasticizers on glassy polymers sensitivity by unraveling the polymer-plasticizer transient response mechanism and identifying the accurate plasticizer for superior BTEX sensing. This work demonstrates that the incorporation of suitable plasticizers into a polymer-based sensor can vastly improve the selectivity for BTEX compounds based on properties such as sorption (via thermodynamic partition coefficients), saturation dynamics, stability, and viscoelasticity of the film. A mathematical model will be incorporated to select the optimum polymer-plasticizer blend for BTEX sensing in gas and liquid phase using QCM and SAW devices.

575123 Microfabricated Silion Nanowire Array Platform for Surface Enhanced Raman Spectroscopy

This work presents a microfabricated silicon nanowire array chip for Surface Enhanced Raman Spectroscopy (SERS) to rapidly detect trace target analytes in biofluids such as serum, urine and exhaled breath. Silicon nanowire array are created by wet etching and coated with silver nanoparticle. Silver nanoparticle coated silicon nanowire array chips are characterized for surface enhanced Raman spectroscopy (SERS). The silicon nanowire array can result in significant SERS for target analyst. The chips are used to detect trace THC in exhaled breath by SERS. The results show that silicon nanowire array coated with sputtered silver nanoparticles can achieve a limit of detection of 3.1 pg of THC. The linear relationship between SERS signal and the amount of THC indicate that the method is suitable for quantification of THC concentration in breath.

AUTHOR INDEX

AUTHOR INDEX

AUTHOR INDEX

AIChE
120 Wall Street, FL 23
New York, NY 10005-4020

ISBN 978-1-7138-0558-8

CURRAN ASSOCIATES INC.
proceedings.com